Use and Abuse
of
America's Natural Resources

Use and Abuse
of
America's Natural Resources

Advisory Editor

STUART BRUCHEY
Allan Nevins Professor of American
Economic History, Columbia University

Associate Editor

ELEANOR BRUCHEY

THE BIRTH OF THE
OIL INDUSTRY

By Paul H. Giddens

ARNO PRESS

A NEW YORK TIMES COMPANY

New York • 1972

Reprint Edition 1972 by Arno Press Inc.

Reprinted from a copy in The Princeton
University Library

Use and Abuse of America's Natural Resources
ISBN for complete set: 0-405-04500-X
See last pages of this volume for titles.

Manufactured in the United States

Library of Congress Cataloging in Publication Data

Giddens, Paul Henry, 1903-
    The birth of the oil industry.

    (Use and abuse of America's natural resources)
    Reprint of the 1938 ed.
    Bibliography: p.
    1. Petroleum industry and trade--United States.
I. Title. II. Series.
TN872.A5G45   1972   338.2'7'2820973   72-2839
ISBN 0-405-04507-7

1763492

THE BIRTH OF THE OIL INDUSTRY

THE DRAKE WELL IN 1866

Peter Wilson and E. L. Drake are shown
standing in the foreground.

THE BIRTH OF THE
OIL INDUSTRY

By Paul H. Giddens
Professor of History and Political Science
Allegheny College

INTRODUCTION
By Ida M. Tarbell
Author of "History of the Standard
Oil Company"

THE MACMILLAN COMPANY
New York — 1938

PRINTED IN THE UNITED STATES OF AMERICA
NORWOOD PRESS — NORWOOD, MASS., U.S.A.

To
MARIE

PREFACE

EDWIN C. BELL, a twenty-three-year-old Yankee, came to the oil fields of western Pennsylvania about 1868 and engaged in printing and newspaper work in Titusville, Pennsylvania. Fully alive to the swift and dramatic developments of the petroleum business and possessing a keen appreciation of its historical importance, Mr. Bell began, and continued for the next fifty years, his acquisition of all sorts of materials relating to the petroleum industry, which formed the nucleus of the present collection at the Museum of the Drake Well Memorial Park, near Titusville.

As the collection at the Museum grew, and as valuable data, heretofore unknown, threw new light on the early days in the northwestern Pennsylvania oil field, a group of interested persons believed that there ought to be a reliable and readable account of the beginnings of the petroleum industry. Although many contemporary books, periodical articles, and newspaper stories had been written on the subject, no recent attempt had been made to present a clear, concise, and connected story of these formative years. The present study, covering the early history to 1870, is, therefore, an outgrowth of this desire.

A large number of librarians, newspaper editors, oilmen, state officials, local historians, and good friends, whose names cannot be mentioned here for lack of space, have generously aided the writer in a variety of ways. Mrs. Elisabeth F. Hequembourg of Titusville has contributed much from her personal knowledge and unique collection of historical mate-

rials. Mr. T. S. Bogardus, Assistant City Engineer of Meadville, has prepared the different maps of the oil region. Miss
Edith Rowley, Librarian, and the staff of the William Edward Reis Library of Allegheny College have been most help-
ful in locating and securing necessary reference works. I am
very grateful to Professor Julian Ross of Allegheny College
for reading the manuscript and proof and for making many
constructive suggestions. I am especially indebted to Miss
Ida M. Tarbell for her constant, wise, and able counsel at
all stages; to Mr. S. Y. Ramage of Oil City, Mr. John H.
Scheide and Mr. James H. Caldwell of Titusville, and the
group headed by them which initiated and made possible
this project, for their invaluable assistance from the beginning; and finally, to President William Pearson Tolley and
the Board of Trustees of Allegheny College for releasing me
on two different occasions from regular academic duties in
order to make this study.

PAUL H. GIDDENS

MEADVILLE, PENNSYLVANIA
August 15, 1938

CONTENTS

ILLUSTRATIONS

MAPS

INTRODUCTION

Dr. Giddens's detailed and graphic study of the Birth of the Oil Industry has particular value as a working model of the American industrial system at the middle period of our national life. Here we see what we were ready to do, what we did, when faced with the possibility that a scarce and isolated natural resource already known to be capable of giving greatly needed service to mankind might be found in quantity.

Here we have in free and active operation all the scientific, financial, mechanical, marketing agencies then existing for industrial development. Here we have demonstrations of the enterprise and resourcefulness of American men in adapting what they knew to unheard-of industrial problems, of their patience and imagination in adding by invention, by trial and error, a body of entirely new mechanical and commercial devices and processes.

Most important, we have here a cross-section of the types of men who had grown up under the functioning of the democratic experiment allowing individual freedom of action. Given full play in the ten years covered by this study we have demonstrations of the evil as well as the good inherent in a system under which these men worked practically without regulation other than their own notions of honor and fair play.

As Dr. Giddens proves at the start, the development of the American oil industry does not begin as is commonly said with the "discovery" of oil in August, 1859, near Titusville,

Pennsylvania. Dating a discovery, like dating the beginning of a war or a revolution, is one of history's most misleading short cuts. Discoveries, like wars and revolutions, have long backgrounds. For two hundred or more years oil had been known to exist in the quarter where the proof that it could be obtained in quantities was finally found.

As early as 1627 a French missionary reported a *fontaine de bitume,* or oil spring, near Cuba, New York. A well known path of the Seneca Indians led southward from that territory along a stream which, at a point we now know as Oil City, emptied into a river the Indians called Allegheny. Close to the banks of this stream were several big oil springs. There were evidences in their neighborhood that the Indians, or possibly a preceding race, had worked them, and collected oil from pits walled by logs.

The white men in their explorations followed the paths of the Indians. Missionaries, hawkers, walkers, travellers, military detachments of the French and English armies, all of the advance guard of settlement, travelled these paths along the Allegheny River and its side streams. Many of them went home to tell of the oil they had seen on their travels, of its uses by the Indians as a medicine.

From these records and tales a map was made in London in 1755 on which the word "petroleum" was written across the territory in which the oil springs occurred. At the point where the stream entered the Allegheny River was the word "Wenango." Thirty-six years later, in 1791, came a map of Pennsylvania on which the stream is named for the first time, "Oyl Creek." From this time on, all gazetteers with a map of Pennsylvania included Oil Creek.

That is, when the nineteenth century opened it was known in this country as well as in England and France that in northwestern Pennsylvania and southwestern New York a substance called rock oil, or petroleum, flowed from springs. It had medicinal value, burned freely, giving a strong but smoky light, was useful as a grease; but its odor was so offensive that many doubted it would ever be widely used for either luminant or lubricant.

When the nineteenth century opened there was something going on in the region more important to the future oil industry than the reports of casual travellers, missionaries, and explorers. It was the steady settlement of the valley of Oil Creek, as well as of adjacent territory, with substantial men and women largely from New England and northern and central New York State. One of the first of these came in 1796— Jonathan Titus. He was soon joined by other settlers. Nine years after they broke ground they plotted a town, which later became Titusville. A few years later the town was laid out according to this plot, laid out as it is today even to the very names of streets.

Titusville was by no means the only settlement in the territory that flourished in these years. In every direction up and down Oil Creek, east and west, along French Creek, the first valley of importance to the west, settlement went on. In all directions were farms in various stages of development. Particularly important to the future of the Region were the farms along Oil Creek.

The rapidity and the solidity of this early settlement were due in a degree to the activities of the agents of the Holland Land Company operating in near-by territory. Originally this Company had laid claim to all western New York and northwestern Pennsylvania; but the laws of the latter state had forced it to whittle down its territory to scattered tracts in six or eight of the northwest counties. These tracts it held by purchase from the state. About the time Jonathan Titus settled on Oil Creek the Holland Land Company sent to Mead's, a settlement thirty miles to the west, an ambitious agent of a distinguished New England family—Major John Alden.

Parallel with the material development went on a substantial cultural development. The Holland Land agents had sought not only woodsmen, farmers, craftsmen, for their enterprise, but educators. John Alden talked of a college, and in 1815 a kinsman of his, Timothy Alden, a scholar well known in the East, worked his way overland to establish what was to be the first college of the district—Allegheny College.

There came in the same year to the region of Titusville from a home in Connecticut a cultivated clergyman, Amos Chase, a man of fine antecedents with the highest outlook on life. For twenty years Amos Chase travelled the region in and around Titusville, establishing churches. Before his death he had founded some thirty-two.

The Chases and others of their kind brought with them the ambition for good and solid homes. There are to be found in and around the Region today many evidences of the influence they had on the building in the first half of the century: fine doorways, moldings, mantels, well proportioned rooms.

From their first coming the settlers in and around Titusville had been alert to the commercial value of the curious product which showed itself in the oil springs of the valley. They had not been long in the country before they discovered that the output of these strange springs was marketable. Buyers came from Pittsburgh on foot or on horseback and took away all they could carry. They discovered that its chief value was for medicine—Seneca oil, it was called after the Indians of the section. They saw the trade grow until all they could dip from their springs was easily sold at a dollar a gallon.

It was not only for medicine that it began to be talked of. Its burning qualities had been known from the first. The odor and the smoke made it impractical save perhaps for outdoor torches. But even if it could be refined into a good luminant, as some men said (Samuel Kier, the head of the biggest medicine trade and one of their best customers from

Pittsburgh, said that he was doing it)—even if that was so, there was not enough of it to be of much importance. Or was there? How were they going to find out? The way they did find out in spite of all difficulties is a graphic illustration of the process by which natural resources were developing in the United States at this time.

These men who settled the new region had not been slow to let their neighbors back home in Vermont, Connecticut, and northern and central New York know about their land, to tell them about the great American medicine and sometimes send them samples to try on their rheumatism, their coughs and colds. These letters brought an occasional visitor curious to see what their friends had found.

South of Titusville on the land of Brewer, Watson and Company, a lumber enterprise, was one of the largest of the oil springs. Every now and then the firm made a nice little sum by selling its oil. Ebenezer Brewer of the firm had a son back in Vermont, an educated physician, to whom he sent samples of rock oil. Dr. Brewer tried it in his practice, found it useful, came on to see his father and look at the springs. Out of that visit grew a series of happenings which were to lead in the next eight years to the founding of the oil industry. They began when Dr. Brewer (who in 1851 had moved to Titusville) took a sample of oil to the laboratory of Dartmouth College for analysis. The reports were enthusiastic. It was all a question of quantity.

While the chemist at Dartmouth College was extolling the possibilities in petroleum an alumnus dropped in, George H. Bissell, a keen curious mind, willing to try things out. Bissell saw at once the commercial possibilities if there were reservoirs of oil below the surface which would yield largely on being tapped. There was no way of knowing except by testing. That took money. There was little delay. With a partner, Jonathan G. Eveleth, he organized the first oil company in the history of the industry—the Pennsylvania Rock

Oil Company—and began peddling its stock. But it was slow work. Times were hard in the fifties. A minor panic in '54 put twenty thousand men in the bread lines in New York City; a major one in '57 a still larger number. A long depression followed. Speculation in stocks which had no substantial backing had been largely responsible for the distress. Men were for the moment cautious about investing money.

Again, oil had not demonstrated itself to the public as had gold and silver, iron ore and coal. It was a new and problematical product. After five years of organizing and reorganizing, of experimenting and failure, an oil company under a new name, the Seneca Oil Company, finally had enough money on hand or in sight to delegate a man to test the land it controlled (a parcel leased from Brewer, Watson and Company) for the presence of oil below the surface in quantities sufficient to justify development on a large scale.

The man sent out in May, 1858, to direct the search had come as accidentally into the undertaking as all the others concerned. He was a railroad conductor, a Jack-of-all-trades— "Colonel" E. L. Drake.

The charter under which the Seneca Oil Company operated allowed digging, boring, or mining for oil. With perhaps three exceptions every man concerned in the venture believed that it was by digging that oil would be found. Was not that the way it had always been found? Their conviction was a demonstration of the blindness of men to the meaning of that which is every day before their eyes. For years advertisements of Seneca Oil in drugstores, as well as newspapers, had carried a picture and a legend explaining the source of the wonderful cure-all. It had been discovered, so ran the legend, in boring for salt water, *not* digging. And there was a picture of the derrick and engine house which carried the drilling machinery.

The first man so far as we know to see the meaning of this advertisement of Seneca Oil was Bissell, the promoter who had seized on the Dartmouth College laboratory reports and had determined to find out for himself where oil came from —how much of it could be obtained. While he was loitering before a drugstore window his eye fell on a bottle of Seneca Oil. Instantly he took the hint. It would be by drilling they would find oil if it was to be in quantity. The President of the Seneca Oil Company, James M. Townsend of New Haven, Connecticut, adopted the theory or worked it out for himself. Probably it was he who converted Drake, for Drake after digging decided to drill.

No invention was necessary. In the fifty years before his attempt the process and all the equipment for putting down a well had been laboriously worked out in seeking salt water. Drake knew he must have an engine and a boiler and drilling tools. He knew that these tools consisted of a chisel bit, sucker rods, jars; that he would need seed bags, a sand pump; that he must have a derrick and an enginehouse in which to set up this equipment. He also must have a man who knew how to use it, a driller. It took time and patience to find them and bring them to the site where he meant to put down the well. The lookers-on of the countryside did not make it easy for him. There was a general inclination to call the undertaking folly, but he had the will to go ahead with the method which he believed to be right, and he stuck to his drilling until the question was answered as to whether you would not find petroleum in quantity by boring into the earth, for on August 27, 1859, at sixty-nine and a half feet, the hole he had made filled with oil. In the first twenty-four hours they pumped probably twenty-five barrels. Wild rumors ran that it was many times that. At the moment the Drake well filled up oil was selling at forty cents a gallon.

The news that Drake had struck oil was a signal for a rush such as the country had not seen since the gold rush of '49. It began in Titusville. Many a man who had jeered at the idea of drilling for oil was quick to spring into action.

Up and down the creek men went buying and leasing land. They did not wait as Drake had done for engines and tools. They improvised tools and put down their first well with the hickory drill, as they called it, a hickory pole worked by foot power. As quickly as men sprang to drilling, others hurried to furnish equipment: blacksmiths turned toolmakers overnight; carpenters fell to putting up enginehouses and derricks. There was no lack of men for any needed service, for the news had gone out overnight by grapevine through the country for fifty miles around and men picked up their bags and tools, and if they had any money, which was doubtful, put it in an inside pocket. They came in by foot, by horseback, sometimes by team, to see what was going on, and what there might be for them. Ready for anything, many of them never left; they had found their chance.

Letters began to go out, letters back home where men were feeling the pinch of hard times. "Come on, greater than California," they urged. Reports began to reach the eastern coast in the Press. On September 12th, sixteen days after the well filled up, the *New York Tribune* published a letter describing what had happened. "The excitement of the discovery of this vast source of oil," said its correspondent, "was fully equal to what I ever saw in California when a large lump of gold was accidentally turned out."

The news reached a population naturally eager for adventure, doubly eager after the fifties' long period of hard times. They flocked from all directions. One of the great advantages of the Oil Region was that it was accessible to population. It lay at a juncture of corners of our three most populous states—New York, Pennsylvania, and Ohio. Close to a third of the country's population—then about thirty-one and a half million—was found in these three states. In them, too, were the chief distributing points. Moreover, the Oil Region lay within comparatively few miles of the greatest railroads of the times—the New York Central, the Pennsylvania, the Erie, and the Baltimore and Ohio. To be sure it meant sixteen, twenty, thirty miles by stagecoach, team, horseback or very often by foot over roads which from the start and throughout our period, were a constant source of cursing. But at least you did not have to cross the continent to find the heart of the excitement. This comparative accessibility, together with the nearness of trunk lines from East to West and the Allegheny River on which the settlers had long chiefly depended, accounts largely for the rapidity of the inrush.

No one in this invading crowd, no one of those who had long lived close to the oil springs knew what to expect from the drilling soon under way. Drake had found oil at sixty-nine and a half feet close to an oil spring and in a first sand. Did it always make itself known by coming to the surface? Could it be found at greater depths? What was the nature of the deposit? A reservoir? A vein? A running stream? These were the questions that overnight men set out to settle.

Every month of every year of the period we are following contributed something to the answer, added something to the unknown geology of oil. The second well, no great distance from the first, put down by a hickory drill, found no oil in the first sand; but drilling a hundred and fifty feet farther, it struck a second sand yielding twenty barrels a day. A month later a third well near by came in pumping seventy to eighty barrels. By the end of 1860 seventy-four wells had been put down, and altogether a half-million barrels had been produced—enough to give better lights to hundreds of homes, enough to convince even the most skeptical that there might be larger and more enduring reservoirs of oil than they had dared dream.

In the second year of the development came a startling and amazing phenomenon—the flowing well. The oil burst forth in powerful gas-driven geysers, often higher than the derrick, sometimes carrying along the drilling tools. What to do with a runaway well? How harness it?

But a more dreadful phenomenon came with the flowing well, the danger of explosion in a gas-filled atmosphere, of fire in an oil-drenched location. The first ghastly demonstration of this danger came on April 17, 1861. A well was drilling three miles from the mouth of Oil Creek. It suddenly began flowing at the estimated rate of three thousand barrels a day. Immense quantities of gas came with the oil; the gas spread to a burning lamp or a lighted cigar, and instantly there was a terrible explosion. It left nineteen dead bodies on the ground and a geyser of burning oil lighting the gruesome spot.

Difficult and dangerous to handle as the fountain, or flowing, well showed itself to be, it became on the instant the ambition of every producer. By the end of 1862 there were seventy-five flowing wells in the Oil Region. With the story of the big fortunes they were making for the lucky owners running wild over the country, it was little wonder that there was a tremendous effort to get in on this good thing. By the end of the first five years of the development over five hundred forty-three oil companies had been organized with a capital close to three hundred sixty million dollars. All of these were organized without the promoters' knowing whether there was oil under the land they proposed to drill, and some of them did not even have leases; but the delirious public was not stopping to investigate what it was buying. One company in New York closed its books four hours after they were opened.

Two types of men from the beginning led in the development of the oil country: those who took it as a speculation, and those who saw in it the possibility of a stable industry. While the get-rich-quick promoter took his winnings and hurried back to the "States" (his way of saying that the Oil Region was no place for a civilized man), those who saw an industry in the making settled in the oil towns. It was the latter who set their minds at once to developing more orderly procedure in the industry, who encouraged the army of mechanical men, sometimes of excellent engineering training who had come into the country to try their luck. They demanded better drills, better engines, better handling of wells, and got them.

It was they who laid the foundation of oil geology. At the start there was no body of facts on which to base a scientific approach to their problem. All they could do was to study and report, record the various layers of rock and sand through which the drill made its way. It was not long before the contents of every sand pump was examined and the nature and thickness of each particular stratum recorded. When oil was found, the characteristics of the sand were noted and compared with those of other oil sands. Soon every producer went around with a little vial of oil sand in his pocket; and the more informed and serious ones carried charts on which were recorded the different kinds of material that had led to their oil sand. That is, each well became a geological study. Before the end of this period the state geologists were in the field, and in 1875 an official report of their findings was published.

There was rapid multiplication of detailed and practical information. Maps began early to be made. By 1865 practically every oil farm was plotted, and every well then in existence marked. Every oilman carried one of these maps in his breast pocket, filling it out as news came of some new strike. The newspapers soon appeared, and luckily they prided themselves not only on keeping up with developments but on trustworthy reporting. It was due to the care of the newspapers that the daily production of the various fields was finally known, that new wells were reported at something like their true yield. In the early years nobody knew—and it was the habit of the country to make every well a gusher. The production of the Drake well was quoted at eighty-two thousand barrels for 1859; but the United States Bureau of Mines puts it at two thousand.

In no particular is the character of the men struggling to put a solid bottom under the new industry better shown than in their concern for the character of the towns in which they were settling. Dr. Giddens's spirited narrative of Titusville in these ten years shows a town fighting for its soul and helped in its struggle by the best element of the incoming crowd.

It is worth noting that oil did not make Titusville. The town was there when Drake came. Its standards of conduct were fixed; it believed in getting on by hard and honest work. Its ambition for education and cultivation was alive.

Then came the deluge bringing with it every form of vice. The town rose to control it and succeeded. But Titusville was not alone in this fight. Six miles away was the little town of Pleasantville, settled thirty years earlier by the colonists of the Holland Land Company. In 1868 Pleasantville had its oil boom beginning with a fourth sand well, located and drilled under spiritualistic direction. Pleasantville was a small civilized community, and vice was never able to get a foothold within the town's limits.

The experience of both of these towns is a demonstration of the impregnability of the standards their people had established in the years before oil was found under their soil. It is worth noting that the leading families, the solid citizens who keep the towns going today, are mainly the descendants of these early settlers, or of those who joined them from the outside when the oil development began.

In sharp contrast to the methods of handling oil fields and of building towns employed by men who sought growth and stability for the new industry was the gambling technique of the invading speculators. Disastrous as this technique often was to the Oil Region, the gambler's scramble to try his luck was a large factor in the rapid geographical spread of the region, proving that oil could be found beyond the banks of Oil Creek and the Allegheny River.

Oil Creek flowed through a valley of varying widths according to the ease with which the strata had been eroded. It was wide and generous at Titusville, narrow and rough below. The erosion which had made the valley had cut its steep sides into gorges down which tumbled runs, as they were usually called, into Oil Creek. Not far from that first tragic flowing well one of these little eroded gorges joined the main valley—Cherry Run valley. Was it possible that oil might be found along the tributaries of the main stream? There was doubt about it, but there was always a chance; and various individuals and companies had early taken up land there. In 1864 came a startling demonstration of the possibilities of oil up the tributary gorges. A man with little money had leased an acre and started a well; a partner joined him. They went down seven hundred feet—a long distance for that day—without oil. But on they went, trusting and drilling, and they soon had their reward, a three-hundred-barrel well.

It was a spectacular demonstration of what might be beyond the limits the first oilmen had set. You did not have to hug the big streams. Overnight scores of operators along the Creek whose wells had passed the flowing stage and settled down to a modest yield shut down their pumps and rushed to the new region. It was not long before there was not a foot of land free from end to end of Cherry Run. Derricks crowded one another on the lowland and up the steep sides of hills. And it was these wells drilling down through the hills which proved the early oilmen wrong who declared oil was to be found only on lowlands. It was a new bit of knowledge which was to be of vast importance in the future.

Many a big fortune was made in the new field by selling a gamble to some more daring gambler. Reed, who drilled the first well on his acre in Cherry Run valley, after selling $75,000 of oil sold his interest for $200,000. A farm of fifty acres next to his well which in '63 had passed hands for a yoke of oxen sold in '65 for $650,000.

The success in Cherry Run greatly stimulated drilling along the tributaries, not only of Oil Creek but of the Allegheny River; French Creek valley to the west of Oil City and one of its tributaries, Sugar Creek, were showing up handsomely. There were tributaries to the east which had not been tested. The first of these was the Pithole Creek. As early as 1863 its possibilities had been discussed and leasing had begun. Several companies were soon in the field with money to develop. The first to go into action was the United States Petroleum Company. Locating a well with a witch-hazel twig, it brought in a gusher in January, 1865, which started at two hundred fifty barrels and gained daily. "The biggest well in the world," the excited drillers named it.

News of the United States well brought men from the length and breadth of the Oil Region. They left wells unfinished, tools in the hole, and hurried to a field which for eighteen months was to prove the maddest gamble so far in the history of the Pennsylvania Oil Region—Pithole. It is said that the rush in a few weeks turned a community of seven persons into one of about fifteen thousand.

The rise and fall of Pithole illustrate what happens when the gambler with his passion for quick returns has the upper hand in a new industrial adventure.

These invading producers and gamblers had a new and continued stimulus, for, while so far in the development wells had usually settled down to steady and often small productions, the United States held up. Six months after it first came in, it was flowing twelve hundred barrels a day. Neighboring wells performed in the same way, kept up their production. Could it be that there were inexhaustible reservoirs?

A city sprang up at Pithole: a city of shanties and saloons which with incredible swiftness became a city of great hotels, fine homes, theaters, an organized town government, a police force, a fire brigade. In a few months the post office of Pithole was handling five thousand pieces of mail a day. The town was full of disbanded soldiers. Distinguished visitors were coming, even British capitalists. Never had buying and selling been keener, bigger prices offered and taken. A farm 'which had brought $25,000 in January sold in September for $2,000,000. There was no end to the riches in sight, they believed.

Then suddenly there came a blow full in the face. First one and then a second of the gushers which had been giving more oil daily rather than less, without warning stopped flowing. There was a big vein of superstition in early oil producers. Trust your future to the word of a medium, the twitching of a hazel rod, and they bring you luck; naturally you watch for signs—heed them. What did this sudden stoppage mean? Was it not a warning that the end of their riches was near? When in the tenth month of its life the United States well abruptly failed panic seized the speculative operators.

What had happened was no mystery. The wells so suddenly dry had been flooded with water, an inevitable result of hasty drilling, of failure to protect the hole by casing and seed-bagging. It had been already demonstrated that wells could be kept active by these means, and there were fields in operation where the failure to follow the practice was considered a reason for cancelling a lease.

Certain of the Pithole operators now combined to make the safeguards obligatory in the field, but the heart had gone out of the boom. Fields were discontinued, and a year from its beginning Pithole had lost half of its population, fires were ravaging its streets, banks were failing, the great hotels and buildings were being raffled off and salvaged. All of this despair was helped by the nation-wide depression of 1866.

But the decline of Pithole by no means put an end to oil operation. Indeed, new fields were opening while eyes were fixed on Pithole, several of them coming into notice in the disastrous year of 1866. There was Petroleum Centre on the Creek, midway between the Drake well and Oil City, which began to boom as the bottom began to fall out of Pithole. Here wells were drilled on the hills and came in big, putting an end for good and all to the curious doubt of oil being under the hills.

And while the development went on over the hills, at the same time new fields were opening up and down the Oil Creek and its tributaries, up and down the Allegheny River, northeast, southwest. New territories were opening in Warren County, down the river at Parker's Landing and Butler, north around Bradford, south in West Virginia—all promising oil in large quantities. The country was learning, too, through the Federal State Department, the extent of foreign oil fields, and oilmen were trying their luck in them. Bissell, the man who had launched the Pennsylvania Rock Oil Company, now helped start both a producing and a refining company in Peru. Robert Locke, who as a boy worked around the Drake well and who still lives in Titusville, in 1877 drilled a well for the Chinese government among the headhunters of Formosa.

This activity enormously increased production in and without the United States. In 1860 five hundred and nine thousand barrels of oil were produced in the United States, five thousand outside. In 1870 the comparative yields were five million, two hundred sixty-one thousand, and five hundred thirty-eight thousand. From a percentage of 98.2 in the world's production we had fallen in ten years to 90.7.

While the oilmen were feeling out the underground pools and channels of petroleum, charting the rocks in which it was found, developing improved methods of drilling and of handling the succession of unexpected problems which striking oil raised, they were also developing two subsidiary industries which were to be of enormous importance in the future of the product. These were the storing and the carrying of oil—forerunners of the tank farm and the interstate pipe line of today.

Apparently the first question Colonel Drake asked himself after he began to pump oil was what he was going to put it in. He had not thought so far ahead. A wash tub is said to have been the first thing he laid his hands on. Barrels of course were the solution, but where were they to come from? Town and countryside fell to making them. Barrel makers in all the towns within a hundred miles doubled and quadrupled their output. The coopers multiplied their plants, but the inadequacy was speedily demonstrated. In the first place they had to be taken away from the well for shipment. When they had all been filled where was the oil to go? The resourcefulness of the men on the ground, the energy with which they met needs, at once seized on this problem. It was answered by a wooden tank, calked with oakum, held together with iron hoops. This appeared in the spring of 1861. Tanks could be made to hold five hundred, a thousand barrels, but not ten thousand.

It was then that the iron tank came in to solve the problem. We had begun to roll iron plates in the United States some thirty years before Drake drilled his well. They had been used early for cylindrical containers. Robert Fulton made the boiler for the "Rocket" from iron plates, riveted together. As soon as the Oil Region realized what had been let loose it turned to the iron tank.

By 1868 there were some eight hundred thousand barrels of iron tankage in the Oil Region. That is, at this early day the oil fraternity had solved the problem of caring for surplus production, of holding the output of fat years for use in the lean ones. The practice of storing oil so early undertaken by producers was to become one of the stabilizing factors in the industry, as well as one of the elements of that control which already different factors in the industry were attempting—the control of price.

The iron tanks would have been less successful in solving the problem of handling the increasing quantities of oil if iron tank cars had not followed them. For five years oil went out of the Region in barrels piled high on flatcars. Then a clever mind suggested that mounting two wooden tanks firmly fastened to the platform of the flatcars would be an improvement. But they soon were to do better than that. An iron cylinder as long as a flatcar made like the iron tank from riveted plates took the place. It held more, was easier to handle, was safer in case of accident. By the end of our period it was the standard shipping car, as it still is.

Along with the search for adequate containers went that for adequate transportation. At the outset everything—lumber, machinery, pipe, barrels, oil—must come in and go out by team. They could ship by water from Oil Creek into the Allegheny River, but the team must carry the shipment to and from the flatboats. They could bring in and send out by rail; but to reach the rail they must drive sixteen, twenty, thirty miles, over roads which for a good part of the year were all but impassable. As production increased, the teamsters multiplied; as they became more and more important, their truculence and extortion doubled. They were endured because indispensable. Their difficult work was met with far more curses than appreciation.

The coming in of great wells in new fields, the rise of boom towns like Pithole created up to 1866 almost unbearable transportation situations. The first prospect of release came from one of a class to which the producer charged much of his trouble—the oil buyer. These buyers took the oil from the well and were far less patient with the teamster than the oilmen who had grown up in the Region and knew by experience the difficulties of the business. In 1865 one of the fraternity in Pithole conceived the idea of making himself the only buyer at that point by cornering barrels. But with his corner fairly complete he found himself at the mercy of the teamsters, a class which had no interest in any monopoly but its own. This man, Samuel Van Syckel, decided that by laying a pipe line from Pithole to the nearest shipping point ten miles away he would be free of the tyrannical element in his business.

Van Syckel had seen and heard enough of what had already been done with pipe lines to believe that he could do something more ambitious. His proposition was met with threats and ridicule. He went ahead determinedly, but when he came actually to laying the pipes the teamsters made good their threat and tore them up as fast as they were laid. Finally Van Syckel was obliged to set up armed guards to protect his enterprise. In the end he was victorious. The line was laid, oil actually began to flow. The whole Oil Region, convinced by what Van Syckel had done, fell to laying longer and longer lines.

The success of the pipe line was tragic for the teamster. It was the Oil Region's first experience with the destruction of a whole class by technological improvement. There were literally thousands of these men with their teams and wagons; but their day was over, and they must either find other jobs or leave the Region. Not a few of them became producers, some of them very rich producers.

The oil industry was free of its first autocrat, but the very instrument which had freed it soon was a factor in building up a second autocrat which the oilmen began to fear more than they had feared the first. This was a system of direct transportation from well to railroad by the combination and control of storage tanks, pipe lines, and tank cars. The system was put into operation by intelligent oil buyers, long harassed by teamsters, lack of storing facilities, difficulty in getting cars. Rapidly there developed in the Region three main companies controlling pipe lines, owning their storage, and—

most important as it turned out—building and owning tank car lines, each of these tank car lines tied to a particular railroad.

The result of this neat and complete consolidation, its obvious alliance with railroads, was that independent shippers and small buyers soon found it difficult to get cars for their oil. They also suspected and loudly declared that the railroads were giving special rates to these tank-car consolidations. At once there came a loud cry of monopoly in the Region. The pipe line which had rescued them from the teamster threatened to become a stronger tyrant than ever he had been.

Their suspicions were justified. The efficiency and convenience of the new unit in the transportation system gathering the oil at the wells and carrying it rapidly with little loss and much less danger of explosion and fire than under earlier helter-skelter methods led the railroads to accept the tank-car lines as subsidiaries, to strengthen the relationship by buying their stock. They looked to control by ownership if necessary.

Behind the spectacular increase of oil production in the first ten years after Drake drilled his well; behind the daring and resourceful men who created and organized revolutionary methods of handling, storing, and transporting the product, were two urgent world needs. The first and more generally recognized was the demand for a better and cheaper luminant. The higher standard of living for poor, as well as rich, of which civilized nations were talking, demanded more light; so did the multiplying activities of industry intent on raising this standard.

If the first thing asked of rock oil was better and cheaper luminants, the second was better and cheaper lubricants. The demand for these was as urgent if less generally recognized. It had come with the rapid extension of railroads and ships, the multiplication of machines of all sorts born of industry's efforts to meet the world's demand for increased comforts and conveniences.

The animal and vegetable oils on which the machine age had so far depended were becoming too expensive, too limited, for the strain put upon them. Forward-looking industrialists and mechanical engineers realized that there must be a new source of supply, and they had been for some time seeking help from chemists. The chemists who analyzed petroleum for the first oil company had said that here was the basis of first-class lubricants, and that if it could be found in quantity one of the crying needs of industry would be answered. The Drake well had given the answer, though its full meaning was not realized until, in the late seventies and early eighties, lubricating engineering came to be a recognized part of mechanical engineering.

The rapid growth of these demands convinced the producer of crude oil that he could sell at a fair price all he could find—if not in one year, certainly over a reasonable period. If he had more oil on his hands than he could properly sell at a given moment, he could store it and wait until the demand caught up. One obstacle in his way, however, was the man who refined his oil. The two were natural competitors, one wanting to sell high, the other to buy low. Moreover, both of them were more or less under the influence of the stupidest of economic obsessions, one that has worked so much mischief in American industry, the notion that prosperity comes from making products scarce and dear. From the start both refiner and producer often sought to give an impression of scarcity in order to boost prices.

However, if these two primary factors had been free to work out the problem of price they might have fixed a fair standard margin between crude and refined. But they were not free. There had early appeared in the Oil Region a middleman, a buyer or trader, who took the oil at the wells and sold it where he found a market. You found him everywhere, at the wells taking the whole output, on the street corners trading with producer or refiner, on the trains, in the bar-rooms of hotels, buying low, selling high. He was particularly cunning in the use of rumors of new fields, convincing sellers that if they did not take what was offered they must take less in the end. It was not infrequent for rumors to be manufactured to help break both the crude and the refined markets. These rumors might be speedily disproved, but in the meantime they had done their mischief.

The buyers were leaders in forming the transportation companies owning storage tanks, pipe lines, and tank cars— agencies to which the railroads quickly allied themselves, and of which certain refiners began early to seek control. Under these conditions corners and rings multiplied, only too often engineered by what a cynical press called "honorable and religious men."

The oil producer, seeing himself at the mercy of the new transportation alliances, sought escape. The most brilliant and thoughtful oil pamphleteer of that time, one John Ponton, argued his case in 1867. Instead of submitting to the low prices and the high rates which monopolists were forcing, the producer should build iron tanks, store his oil, shut down his wells if necessary; but, while that might force the monopolists into good behavior for a time, there would be no change of heart, only a forced good behavior. If he was to be master of his product, he must build a pipe line to the seaboard, and Ponton attempted to demonstrate that this was feasible. Actual threats of carrying out the idea were afloat when Ponton's pamphlet appeared. It was ridiculed loudly, but the faith of the promoters was backed up by so eminent an authority as General Haupt, whose bridges built during the Civil War had made him famous. General Haupt said that it was possible to pump oil over the mountains. That is, the oilmen saw before they were eight years old that, if they would unite, there might be a possible permanent escape from the forces in the industry trying to harness them for their own profit.

But in the middle sixties oil producers were not ready for united action. It meant an interlocking of the factors of the industry they disliked. Let a man be a producer, a refiner, a pipe-line man, stick to his own job, not complicate it by mixing with other men's jobs. It was the philosophy of the pioneer industrialists, but they were soon learning they had no choice. Integration was on their necks, driving them into all sorts of alliances. It was not long before they were agreeing they must hold together. They already had striking exhibits of what they could do when they acted collectively. Improved methods in putting down wells, fixing standards, collecting information and spreading it, had come about by cooperation.

Their most striking success in collective action had been in 1866, when they secured the repeal of the federal tax on crude oil. This tax had first been put on in 1864, $1.00 a barrel. It proved killing for the small producer. Eight thousand wells were said to have been shut down in a few months. Those who had levied the tax had evidently made no survey of the business as a whole, but had taken it for granted that the great majority of wells were big producers—five hundred, a thousand, three or four thousand barrels a day. Too much money. Now three-fourths of the producing wells at the time the tax was applied were under thirty barrels. Oil was down to $2.00 a barrel; a tax of $1.00 a barrel shut them down. The oilmen, feeling that they had a just cause, began to unite in different fields, to pass resolutions and form committees. Finally a committee was sent to Washington, and it did its work so well that in May of 1866 Congress, on the advice of the Ways and Means Committee, repealed the impost. The news raised a crop of banquets in the Oil Region, and at the same time set small wells to pumping again.

It was one of the first effective lessons in the value of unionism. You began to hear on all sides of applying it to other problems, unfair railroad rates, speculative attacks on price, exorbitant royalties, the growing efforts of one factor in the industry to control others, to form monopolies. The feeling grew that oil producing was no longer a matter for an individual, and must be handled collectively if they were to remain independent. An answer to this conviction was the organization in Titusville in February, 1869, of the first Petroleum Producers' Association. The union had several objectives; indeed, its business was the correction and control of whatever was interfering with orderly business and a fair price structure. It was even prepared to deal with its own outstanding abuse—overproduction—and a few months after its organization was attempting to stop the drill and shut down active wells.

A chief concern of the Producers' Association was putting an end to railroad rebates and discriminations. It was mainly due to these abuses that at the end of the decade crude oil sold on the Creek at around $4.00 a barrel and in New York at around $18.00. It was due to them that outside refining points were taking the business away from the Region where the oil was produced. The attack on their supremacy could not go unchallenged. The railroads must be forced to do their duty, and there began to be talk of a Federal Interstate Commerce bill forbidding practices so contrary to the common law, not to say to the democratic ideal.

Great as were the difficulties of the industry, they did not slow down its growth. As a wealth producer it steadily grew; already it was classified high among exports. But the true value of an industry is not measured in dollars. It is measured by its service to the masses of men. In this first ten years of its life the oil industry had become an invaluable servant, giving cheap and abundant light in quarters where light had been barely known in the dark hours of the twenty-four. Rapidly it was going into the remotest parts, not only of this country but of Europe and Asia. Illuminating oil was contributing more in comfort and advantages to the poor man's home than any other development of that time.

But that was by no means all it was doing by 1870. The problem of making a satisfactory lubricant from it had been solved by the chemists at a time when machines were multiplying. Without good, cheap, and abundant lubricating oil they could not function. Petroleum was supplying it. An immense market awaited it.

Many men who were studying the possibilities of the product saw in it the world's future fuel. The gas which accompanied the oil in immense quantities was used even before Drake's time in localities where it had shown itself. Many drillers used it in their boilers; ironmasters around Pittsburgh were experimenting with it. Crude oil was believed to be a possible fuel, waiting only a proper burner; and on that inventors were working. As early as 1867 petroleum was tried as a fuel for ships and for locomotives and engines.

But an industry does not develop alone. It depends on others and, if it grows, carries them along. The stimulus the oil business gave the lumber business cannot be reckoned, nor can that it gave to iron in many different forms. Not only were old and established trades and occupations energized, extended, but new lines of activity were opened by inventors, answering the call for better tools, containers, transportation equipment, stills and refining processes, more practical lamps and burners.

To achieve these results the industry had called upon all of the scientific, mechanical, and commercial forces of that day. It had used men of imagination who dared to risk all they had on the adventure of seeking oil. It had used capital wherever it could be found. It had used the promoter and the speculator. It had called on the chemist to evaluate the products and had set him up a laboratory to enlarge and improve them. It had called on the engineer to apply all known mechanical devices and to find ways to solve the al-

most daily problems that arose in the producing and handling of oil. It had adapted the markets of the world, ex-
porters, importers, salesmanship, to distributing its products.

The way that all these varied activities fell into line,
promptly and automatically organizing themselves, is one
of the most illuminating exhibits the history of our industry
affords, of how things came about under a self-directed, democratic, individualistic system: the degree to which men
who act on "the instant need of things" naturally supple-
ment one another—pull together.

The system had its evils, but they came chiefly from the
lack of character of a percentage of the men involved. One
may say with confidence that the majority of the men who
built the oil industry were fair dealers, self-reliant, too self-
respecting to give anything but a full return for what they
got. We can say with equal confidence that there was a
considerable minority, including some of the ablest men
in the business, who were willing to employ any device,
however unfair, which would strengthen their individual
fortunes. At every point in the development of the industry there were mean little efforts to cheat. There were men
who put false bottoms in their barrels, selling forty instead
of forty-two gallons, who tapped pipe lines, made false reports of production, broke the word-of-mouth contracts which
the Oil Region held sacred. There were refiners who flooded
the market with kerosene so poor and dangerous that foreign buyers revolted in one year, 1872, buying five million
gallons less than in 1871.

Nor was the industry a stranger to violence. Teamsters
in 1865 tore up pipe lines and for many months carried on
a brutal sabotage of the new transportation system. The
pipe line won; but it soon encountered a new enemy, for the
railroad, seeing it as a dangerous competitor, did not hesi-
tate to use the teamsters' tactics. Disputes over leases became
every now and then pitched battles. The grievance com-

mittees of the exchanges were kept busy with charges of unfair dealings. The courts of the oil-producing counties wrestled with the new interpretations forced by new legal problems.

More sinister than these difficulties was the practice of a number of the ablest refiners, producers, and dealers, of secretly securing advantages contrary to what was regarded as sound business ethics—advantages chiefly in their dealings with the railroads, notorious at that time for their defiance of all obligation.

It is certain, however, the development could never have gone on at anything like the speed that it did except under the American system of free opportunity. Men did not wait to ask if they might go into the Oil Region: they went. They did not ask how to put down a well: they quickly took the processes which other men had developed for other purposes and adapted them to their purpose. Each man made his contribution: one knew of a wooden tank and built it; one had seen the cylindrical iron tank and ordered one; another knew that oil would run through pipes, believed it could be pumped uphill if necessary, tried it, proved it. What was true of production was true of refining, of transportation, of marketing. It was a triumph of individualism. Its evils were the evils that come from giving men of all grades of character freedom of action.

Taken as a whole, a truer exhibit of what must be expected of men working without other regulation than that they voluntarily give themselves is not to be found in our industrial history.

IDA M. TARBELL

CHAPTER I

Early History of Petroleum in the United States

MANY people believe that Colonel E. L. Drake was the first to discover oil in the United States. This is not true. For over two centuries before Drake drilled his famous well in northwestern Pennsylvania, petroleum was known to exist in this country. The earliest record of its presence came from the pen of a Franciscan missionary, Father Joseph de la Roche d'Allion, who traveled through what is now western New York in 1627 and saw "some very good oil." It is generally believed that d'Allion's reference was to an oil spring near the present site of Cuba, New York. Several Jesuit missionaries, visiting the region in 1656 and 1657, apparently found the same spring. They reported that it contained a "heavy and thick water, which ignites like brandy and boils in bubbles of flame when fire is applied to it." [1] It was so oily that all the Indians used it to anoint and grease their heads and bodies. Early in the next century both Charlevoix, another Jesuit, and Pouchot, commander of the French forts at Niagara and Lévis, mentioned the spring.

From the early eighteenth century, English colonial officials and missionaries either saw or heard of the spring at Cuba. In the fall of 1700 the Earl of Bellomont, governor of

---

[1] *The Jesuit Relations and Allied Documents* (Reuben Gold Thwaites, ed., Cleveland, 1900), XLIII, 261, 326. This quotation is used by permission of the Burrows Brothers Company, Cleveland, Ohio.

New York, sent His Majesty's chief engineer in America, Colonel Wolfgang W. Romer, to visit the Five Nations, and, among other things, instructed him to "view a well or spring which is eight miles beyond the Sineks furthest Castle, which they have told me blazes up in a flame when a light coale or firebrand is put into it; you will do well to taste the said water, and give me your opinion thereof and bring with you some of it." [2] While holding an Indian conference at Niagara in September, 1767, Sir William Johnson made an entry in his journal to the effect that "Aseushan came in with a quantity of Curious Oyl, taken of the top of the water of some very small Leake, near the Village he belongs to." [3] A year later David Zeisberger, a Moravian missionary, traveling through western New York, noted that there were "oil wells, with the products of which the Senecas carry on trade with Niagara." [4]

No adequate description of the spring at Cuba appeared until 1833, when the distinguished Yale chemist, Professor Benjamin Silliman, Sr., visited the place and recorded his observations. [5] He found the spring located in a marsh. It was a muddy and dirty circular pool about eighteen feet in diameter, without an outlet. Stagnant water filled the spring, and on top of the water was a thin layer of petroleum, giving it a foul appearance. According to Professor Silliman, the people in the vicinity frequently and freely collected the oil by skimming it, like cream from a milk pan. Flat

[2] *Documents Relative to the Colonial History of the State of New York; Procured in Holland, England, and France* (E. B. O'Callaghan, ed., Albany, 1854), IV, 750.

[3] *The Documentary History of the State of New York* (E. B. O'Callaghan, ed., Albany, 1850), II, 510.

[4] David Zeisberger, "The Diaries of Zeisberger Relating to the First Missions in the Ohio Basin," *Ohio Archaeological and Historical Publications*, XXI, 82. This quotation is used by permission of the Ohio State Archaeological and Historical Society, Columbus, Ohio.

[5] S. F. Peckham, "Production, Technology, and Uses of Petroleum and Its Products," *House Miscellaneous Documents*, No. 42, 47 Cong., 2 sess., XIII, part 10, 8.

upon and just under the surface of the water they moved a broad flat board, made thin at one edge like a knife. When a coating of petroleum had covered the board, they scraped it off into a vessel. To purify the oil, they heated it and, while it was hot, strained it through a flannel or woolen cloth. Once the impurities had been removed, the local inhabitants used it for sprains and rheumatism and for sores on their horses. Professor Silliman could not ascertain the amount of petroleum annually obtained, but he believed that "the quantity must be considerable." He was quite positive, however, that very little of the oil then being used in eastern cities came from the Cuba spring.

Soon after the opening of the nineteenth century petroleum began to appear in the salt wells of West Virginia, Ohio, and Kentucky. The first salt well west of the Allegheny Mountains, if not in the United States, to produce petroleum in any quantity was drilled on the Great Kanawha River, a short distance from Charleston, West Virginia. Located on the north side of the river and just in front of "Thoroughfare Gap" was a salt lick or "The Great Buffalo Lick," as it was commonly called. Through Thoroughfare Gap from the north, and from up and down the river, buffalo, elk, and other animals frequently made their way in vast numbers to the salt lick. In 1806 David and Joseph Ruffner decided to ascertain the source of this salt water in order to procure a larger supply and better quality and to manufacture salt on a scale commensurate with the growing needs of the time.[6] To reach the bottom of this mire, through which the salt water flowed, the Ruffners secured a straight, well formed, hollow sycamore tree, having an internal diameter of four feet, and sawed it off square at each end. They set this hollow tree, large end down, in a perpendicular position on the edge of the salt lick, and held it upright by braces on all four sides. Having built a platform about the top upon

6 *Ibid.*, 6–7.

which two men could stand, they erected a swape, or slanting pole, having its fulcrum in a near-by forked post set in the ground. To one end of this swape they attached a large bucket, made from half a whisky barrel, and to the other end a rope to pull down on and raise the bucket.

With one man inside the hollow tree, armed with a pick, shovel, and crowbar, two men on top to empty and return the bucket, and three or four to work the swape, the Ruffners proceeded to dig. After encountering many delays and difficulties they finally reached bedrock sixteen or seventeen feet below the surface. Since the bottom of the tree was square and the surface of the rock uneven, the rush of outside water into the tree became very troublesome. By cutting and trimming first on one side and then the other, they were able to make a fairly smooth joint, after which they resorted to thin wedges, which were driven in wherever they would do the most good. In this way they made the tree sufficiently tight for the water to be bailed out in order to determine whether or not the salt water came up through the rock. The quantity coming up proved to be extremely small, but its strength was greater than previously. Anxious to follow it down, the Ruffners made a long iron drill with a two-and-a-half-inch chisel bit and attached the upper end to a spring pole with a rope. In this way, drilling continued slowly and tediously until January 15, 1808, when the Ruffners were rewarded by an ample flow of strong brine, and drilling ceased.

They were now confronted with the problem of how to get the stronger brine from the bottom of the well, undiluted by weaker brine and fresh water at the top. Undisturbed by the fact that there was no precedent, the Ruffners whittled out of wood two long half-tubes of the proper size. Fitting the edges carefully together, they wrapped the whole from end to end with small twine. With cloths wrapped around the lower end in order to make it as nearly watertight as possible in the two-and-a-half-inch hole, the wooden tube was cautiously pressed down to its place. It answered the purpose perfectly. The brine flowed up freely through the tube into the hollow tree, which had a watertight floor, and the salt water could now be raised by the swape and bucket.

Without any preliminary study, previous experience or training, precedents, steam power, machine shops, skilled mechanics, suitable tools or materials, the Ruffners had drilled and tubed, rigged and worked their salt well. The wonder is not that they required eighteen months or more to complete the well for use, but rather that the well was actually completed under the circumstances.

The Ruffner well soon began to plague its owners by producing petroleum with the brine, but the amount produced is not known. As other wells were drilled they also contained more or less petroleum. Some of the deepest wells are said to have produced as much as twenty-five to fifty barrels a day. Whatever the amount, it was allowed to flow over the top of the salt cisterns into the river. The oil spread over a large surface, and by its beautiful iridescent hues and not very savory odor it could be traced for many miles downstream. For this reason the river received the name of "Old Greasy" by which it was for a long time familiarly known by Kanawha boatmen and others.

In southern Ohio, the Muskingum River valley from Zanesville to Marietta and the valley of Duck Creek were also noted for their salt wells, but they produced little oil. A salt well drilled in 1814 on Duck Creek, about thirty miles north of Marietta, furnished the greatest quantity of oil in the region.[7] It flowed every two to four days, each eruption lasting from three to six hours, which, for the first few years, amounted to thirty to sixty gallons at each period. The discharges grew less frequent and diminished in quantity until in 1833 the well produced only about a barrel a week.

[7] *Ibid.*, 7–8.

Four or five years after this salt well first started producing oil, Martin Beatty drilled a salt well on the Big South Fork of the Cumberland River, twenty-eight miles southeast of Monticello, Kentucky, which spouted petroleum in such quantities that it was abandoned for brine. Farther west in Barren and Cumberland counties other salt wells were drilled, and in many of them petroleum also appeared. The most famous of all the Kentucky wells was the American well, located near Burkesville.[8] In drilling for salt in 1829 the owner struck a vein of pure oil, an incredible amount of which gushed forth at intervals of two to five minutes, each discharge throwing oil twenty-five to thirty feet in the air. It is claimed that the well produced a thousand barrels of oil a day. Oil floated down the Cumberland River, and for miles covered the whole surface. After continuing to flow for three or four weeks, the well subsided. Later, several persons took charge of the well, saved the oil, put it up in bottles, and sold it as "American Medicinal Oil, Burkesville, Kentucky."

While it is of interest to know that these early salt wells produced oil, the important thing is that, especially in West Virginia, a certain method in drilling was developed and varous improvements, like the use of tubing, a "seedbag," and "jars," were gradually introduced, all of which proved valuable later in drilling the Drake well. In addition to making these significant contributions, the Kanawha salt wells educated and sent forth a set of skillful drillers, some of whom were destined to be among the first oil-well drillers along Oil Creek.

However, the greatest source of petroleum prior to 1845 was along Oil Creek, a small stream in northwestern Pennsylvania. The earliest evidence of its presence here was at the time of the old pits, which once covered the land above and below the junction of Oil Creek and Pine Creek, a short

[8] *Ibid.,* 8.

distance south of Titusville, Pennsylvania.[9] All together there were about two thousand of them scattered over this level plain, and others could be found at intervals down the winding valley of Oil Creek. Each one was seven or eight feet long, four to eight feet wide, and six to ten feet deep; in form, they were circular, oblong, oval, and square. Large timbers, stripped of bark, halved at the ends, and squarely cut, were vertically set up around the sides of the pits in order to prevent cave-ins.

According to the story, water seeped into the pits, and within a short time a layer of petroleum from one-third to one-half inch thick formed upon the surface. A piece of woolen or flannel cloth or a blanket was then thrown into the pit. When saturated with petroleum, the cloth was taken out and wrung into some sort of container. In this crude but effective manner a race of people is supposed to have obtained petroleum. From the number of pits and their systematic arrangement, it is possible that they secured a relatively large quantity of oil.

By whom and at what date the pits were excavated is a mystery. Some believe that the American Indians made them long before the appearance of the white man.[10] Averse to manual labor of any kind, the Indians let their women plant corn, build lodges, and do whatever else was necessary, while they roamed the forest, hunting, fishing, and fighting. Therefore, the women must have dug and cribbed the pits—which seems rather unlikely in this case. Moreover, when questioned in later years, surviving Seneca Indians near

[9] For early descriptive accounts of the pits see the following: Timothy Alden, "Antiquities and Curiosities of Western Pennsylvania," *Archaeologia Americana* (Transactions and Collections of the American Antiquarian Society, Worcester, Mass., 1820), I, 308–320; *Crawford Democrat* (Meadville, Pa.), April 1, 1848; Thomas A. Gale, *The Wonder of the Nineteenth Century: Rock Oil in Pennsylvania and Elsewhere* (Erie, 1860), 15; Edwin C. Bell, *Notes on Journalism* (Titusville, 1910), 8.

[10] S. J. M. Eaton, *Petroleum: A History of the Oil Region of Venango County* (Philadelphia, 1866), 50.

Oil Creek could not give any explanation of the origin of the pits.[11]

Others contend that the French scooped them out during their occupation of the trans-Allegheny region.[12] If so, to what market did they send the oil? To Canada? The records of the time do not indicate that any large quantity was sent in that direction. Furthermore, there were no facilities for shipping much to the seaboard for exportation in case of a strong demand. It is reasonable to believe that if the French had undertaken to cultivate the soil or exploit the natural resources, except in a temporary way, the Indians would have resisted. Finally, since the English and French were such bitter rivals for the West, it is strange that the French would dig so many pits, spend so much time along Oil Creek, and employ so many people in getting oil, without erecting a village or fort, the remains of which might now afford some clue as to what race dug the pits.

The prevailing theory is that the pits were excavated by an ancient, civilized people called the Mound Builders, the probable ancestors of the American Indians.[13] Though proof is lacking for a long period of residence, there is no doubt that they visited the trans-Allegheny region and camped on the present site of Franklin, Pennsylvania. Excavations of the pits in the early part of the nineteenth century indicated that they were lined with timbers on which notches had been cut with blunt stone instruments, which led some antiquarians to conclude that they had been dug and cribbed by the Mound Builders. Of greater importance in substantiating this view was the fact that at the time trees two hundred to three hundred years old stood either among the pits, or on the septa, or directly over some of the pits.[14] They were

[11] *Ibid.*, 53.

[12] *Ibid.*, 47, 49; J. C. G. Kennedy, "Preliminary Report on the Eighth Census, 1860," *House Executive Document*, No. 116, 37 Cong., 2 sess., 71.

[13] Eaton, *op. cit.*, 48.

[14] *Crawford Democrat*, April 1, 1848; Titusville *Morning Herald* (Titusville, Pa.), May 3, 1866. In December, 1869, the title of this paper was changed to *Titusville Morning Herald*.

hardly in existence at the time the pits were made, for it is only reasonable to believe that those making the excavations would have started far away from timber or else would have cut down the trees. There was no evidence in the vicinity of Titusville as late as 1847 to show where the timbers for cribbing might have been cut. The whole flat was covered with standing timber, and not a stump remained to show where the axmen had been prior to its visitation by the whites. Since the digging and lining of the pits and the method of getting oil indicated a people with considerable intelligence and ingenuity and, in some degree, the elements of modern civilization, it is possible that the Mound Builders made them. They were sun or fire worshipers and offered sacrifices to the sun god. Under the circumstances, one might very well inquire, Did they make frequent journeys to obtain petroleum to maintain a perpetual fire in their worship?

In the summer of 1934, the State Archaeologist of Pennsylvania opened six of the pits below Titusville. From his examination, he concluded that the holes were made by the Indians to obtain clay, a pure blue clay highly prized for making pottery vessels, which could be found at a depth of less than four feet.[15] Thus a new and fourth theory was added to the confusion that already existed.

Not until about the middle of the eighteenth century does any written record appear to indicate the existence of petroleum in northwestern Pennsylvania. Often cited as the earliest evidence is the fictitious letter to General Montcalm in 1750 from the commandant at Fort Duquesne, in which he discusses a religious ceremony of the Seneca Indians on a small stream which flowed into the Allegheny River a few miles above Fort Venango.[16] At the end of the ceremonial, a torch was applied to the thick, oily surface of the stream,

[15] A letter to the writer from Donald A. Cadzow, State Archaeologist of the Commonwealth of Pennsylvania, dated April 2, 1937.
[16] Gale, *op. cit.*, 64.

causing a great conflagration. A number of factors incline one to believe, however, that this description is a fabrication. The commandant at Duquesne could not have written to Montcalm in 1750, for the latter did not arrive in Canada until 1756. If such a letter had been written, it probably would have been addressed to the Governor of Canada and not to Montcalm, for the Indians were within the Governor's sphere and the fort commanders were directly under him, not under Montcalm.

The earliest and most reliable information about oil in northwestern Pennsylvania is derived from the diary of David Zeisberger, the Moravian missionary who brought the Gospel to the Indians at Goschgoshuenk, a Monsey town at the junction of Tionesta Creek and the Allegheny River, now the approximate site of Tionesta, Pennsylvania. In 1768 four of Zeisberger's party went on a hunting trip down the Allegheny River. On their return Zeisberger reported that "they brought back some oil from the oil-well. There are said to be various such wells in this region. The oil has a very strong odor, and cannot be used with foods. The Indians use it externally as a medicine and it would be possible to use it for lighting. The oil comes out of the ground with the water and then rises to the top, so that it is possible to skim it. The Indians generally try to get that which has just come up, as it has not so pungent an odor. The nature of these oil-wells might well be investigated." [17] No clue is given as to the location of the "oil-well," but possibly it was near the junction of Oil Creek and the Allegheny. Of the oil springs in this section of the country and their use by the Indians, he significantly says: "I have seen three kinds of oil springs—such as have an outlet, such as have none, and such as rose from the bottom of the creeks. From the first, water and oil flow out together, the oil impregnating

[17] Zeisberger, *op. cit.*, 79. This quotation is used by permission of the Ohio State Archaeological and Historical Society, Columbus, Ohio.

the grass and soil; in the second, it gathers on the surface of the water to the depth of the thickness of a finger; from the third it rises to the surface and flows with the current of the creek. The Indian prefers wells without an outlet. From such they first dip the oil that has accumulated; then stir the well, and when the water has settled, fill their kettles with fresh oil, which they purify by boiling. It is used medicinally, as an ointment for toothache, swellings, rheumatism and sprains. Sometimes it is taken internally. It is of a brown color, and can also be used in lamps. It burns well." [18]

The oil also formed an article of local trade during the latter part of the eighteenth century between the Indians and traders; in some old account books kept at Franklin, Pennsylvania, "gallons" and "kegs" of oil were credited to Indians.[19] There were also instances where the Indians sold it to the whites at four guineas a quart.[20] Having been first used by the Seneca Indians and found in their territory, it was commonly known as "Seneca Oil." Some people, however, called it "Genesee Oil," or "Fossil Oil," or "American Oil." Later, men of science simply called it "Petroleum" or "Rock Oil."

Three of the best accounts of oil along Oil Creek were written by General Benjamin Lincoln, Dr. Johann David Schoepf, and General William Irvine during the period of the Confederation. While marching through western Pennsylvania about 1783, General Lincoln's troops halted at a spring on Oil Creek, collected oil, and bathed their joints with it.[21] This gave great relief and freed many of them

---

[18] *The Derrick's Hand-Book of Petroleum; A Complete Chronological and Statistical Review of Petroleum Developments from 1859 to 1899* (Oil City, 1898), I, 6. This quotation is used by permission of The Derrick Publishing Company, Oil City, Pa.

[19] Kennedy, "Preliminary Report on the Eighth Census, 1860," *House Executive Document*, No. 116, 37 Cong., 2 sess., 71–72.

[20] George Henry Loskiel, *History of the Mission of the United Brethren Among the Indians in North America* (Christian Ignatius La Trobe, trans., London, 1794), 118.   [21] *Derrick's Hand-Book*, I, 6.

immediately from rheumatic complaints. They also drank freely of the waters, which acted as a "gentle purge." Although he did not visit the region, Dr. Schoepf, in his tour through the Confederation in 1783 and 1784, heard of a famous spring on Oil Creek, "which is heavily saturated with it [petroleum], and the broad creek is for a long distance covered with the swimming oil." [22] Appointed to explore a tract of land presented by the State of Pennsylvania to its American Revolutionary War soldiers in 1785, General Irvine and his agents examined the oil seepage on Oil Creek; and in his report to President Dickinson, Irvine wrote: "It has hitherto been taken for granted that the water of the creek was impregnated with it [petroleum], as it was found in so many places, but I have found this to be an error, as I examined it carefully and found it issuing out of two places only—these two are about four hundred yards distant from [each] other, and on opposite sides of the Creek. It rises in the bed of the Creek at very low water, in a dry season. I am told it is found without any mixture of water, and is pure oil; it rises, when the creek is high, from the bottom in small globules, when these reach the surface they break and expand to a surprising extent, and the flake varies in color as it expands; at first it appears yellow and purple only, but as the rays of the sun reach it in more directions, the colors appear to multiply into a greater number than can at once be comprehended." [23]

Reports of the oil spring soon led map makers to indicate its presence. The earliest known map to note the presence of petroleum in northwestern Pennsylvania is Lewis Evans' *Map of the Middle British Colonies in America*, published in 1755. Very close to the present site of Titusville and Oil

[22] Johann David Schoepf, *Travels in the Confederation* (Alfred J. Morrison, ed. and trans., Philadelphia, 1911), I, 254. This quotation is used by permission of William J. Campbell, Philadelphia, Pa.

[23] "Report of General William Irvine to John Dickinson, August 17, 1785," *Pennsylvania Archives,* First Series, XI, no. 25, 516.

City, the word "Petroleum" is printed. Peter Kalm's *Map of New England and the Middle Colonies* in 1772 and Thomas Pownall's *Map of the British Colonies in North America* in 1776 also have the word "Petroleum" printed in about the same spot. The map of Pennsylvania published by Reading Howell in 1791 shows a tiny stream in northwestern Pennsylvania flowing southward into the Allegheny which is labeled "Oyl Creek." This is probably the first map to show Oil Creek, whose name was derived from the fact that oil had been found floating on its surface.

Accounts of this natural curiosity, based upon General Lincoln's observations, found their way into Jedediah Morse's *American Universal Geography* in 1789 under the heading "American Natural Curiosities," the *Massachusetts Magazine* in 1791, and Joseph Scott's *United States Gazetteer* in 1795.

As white settlers moved into northwestern Pennsylvania and settled along Oil Creek and its vicinity, they began collecting petroleum from little springs either in the bank or in the actual bed of the stream; or, making an excavation in the low marshy ground, which immediately filled with water and oil, they skimmed off the oil. Whenever they found a spring in the bed of Oil Creek, they constructed a dam of loose stones a little higher than the surface of the water, ten to fifteen feet in diameter, around the place where the oil bubbled up. An eddy was thus created inside the wall which confined the floating oil, while the water flowed out freely between the loose stones.[24] The oil was allowed to

[24] The early methods used to collect oil are described in the following: *Derrick's Hand-Book,* I, 8; Colonel Drake's Account, 4, a manuscript in the Drake Museum, Titusville, Pennsylvania; J. H. Newton, *History of Venango County, Pennsylvania, and Incidentally of Petroleum, Together with Accounts of the Early Settlement and Progress of Each Township, Borough, and Village, with Personal and Biographical Sketches of the Early Settlers, Representative Men, Family Records, etc.* (Columbus, Ohio, 1879), 205; and Fortescue Cuming, *Sketches of a Tour to the Western Country, Through the States of Ohio and Kentucky; a Voyage Down the Ohio and Mississippi Rivers, and a Trip Through the Mississippi Territory and Part of West Florida* (Pittsburgh, 1810), 84.

accumulate for several days, until it became an inch or more deep, when a piece of flannel or woolen cloth or a blanket was spread over the surface to absorb the oil; then it was wrung out by hand into a barrel or some other container. In this manner a gallon and in some cases several gallons could be gathered in a day. Ten to twelve barrels of oil might be collected in a season, depending upon the prevalence of dry weather and low water. The output of most of the springs, however, was insufficient to warrant collecting it as a business.

The spring which yielded the greatest amount of oil and to which most of the late eighteenth and early nineteenth century observers refer was located in the middle of Oil Creek on the Hamilton McClintock farm about three miles above the junction of Oil Creek and the Allegheny.[25] From it twenty to thirty barrels of pure oil could be obtained in a season.[26] The spring was said to be a source of income to the owner, for oil sold as early as 1807 for one to two dollars a gallon.

The local inhabitants valued and used it exclusively as medicine. They found it a singularly beneficial and infallible cure for headaches, toothache, sore breasts, weakness in the stomach, rheumatism, lung diseases, constipation, cuts, burns, bruises, and all human complaints.[27] Almost every family in the region kept a small supply of Seneca Oil for emergencies; for ordinary purposes a pint bottle would last a year. It also came into general use for saddle burns and scratches in horses. Crude oil seemed to be particularly adapted to horse flesh; flies and insects seemed to have an antipathy toward it, and it proved very effective in preventing the de-

---

[25] Thomas Ashe, *Travels in America Performed in 1806 for the Purpose of Exploring the Rivers Allegheny, Monongahela, Ohio, and Mississippi, and Ascertaining the Produce and Condition of Their Banks and Vicinity* (London, 1808), 46.

[26] Newton, *op. cit.*, 647; Gale, *op. cit.*, 15.

[27] Thaddeus Mason Harris, *The Journal of a Tour into the Territory Northwest of the Allegheny Mountains, Made in the Spring of the Year 1803* (Boston, 1805), 46; Cuming, *op. cit.*, 84.

posit of eggs by the blowfly in the wounds of domestic animals during the summer months.[28]

An amusing story is told of two fellows who got two or three barrels of Seneca Oil from Oil Creek, secured a horse and wagon, gathered up what vials and small bottles they could, and started to peddle it as medicine throughout the neighboring communities. On nearing a village or town, the older of the two men would go ahead, stop at the local tavern, drink brandy, sit down to rest, and soon be rolling on the barroom floor in great distress. At this point the portly partner would enter, asking, "What ails the old man?" Receiving no reply the "doctor" would feel his pulse, and then exclaim, "Gentlemen, I have in my wagon at the door, a medicine that will cure the stranger in ten minutes." A dose would be administered, relief would be instantaneous, and every man present then wanted a bottle of oil irrespective of cost.

Possessing more enterprise and initiative than his neighbors, Nathaniel Carey, one of the first settlers on Oil Creek, often collected or purchased oil and peddled it about the country. Carey is said to have introduced petroleum into Pittsburgh, seventy or eighty miles distant, about 1790.[29] His shipment consisted of two five-gallon kegs that were slung on each side of a horse. In exchange for the oil, he secured necessary provisions for his family. Later, raftsmen would bring a barrel or two down the Allegheny on a raft of lumber or logs; but the introduction of so much oil at one time literally flooded the market. It is also important to note that a considerable quantity was annually received in New York City by 1830 and sold to apothecaries.[30]

[28] J. T. Henry, *The Early and Later History of Petroleum, with Authentic Facts in Regard to Its Development in Western Pennsylvania* (Philadelphia, 1873), 24.

[29] *Derrick's Hand-Book*, I, 9; Newton, *op. cit.*, 206; John Earle Reynolds, *In French Creek Valley* (Meadville, 1938), 72–73.

[30] *New York Journal of Commerce*, quoted in Sherman Day, *Historical Collections of the State of Pennsylvania* (Philadelphia, 1843), 250.

While the quantity of petroleum produced and used prior to 1845 is not exactly known, it is safe to conclude that the amount was insufficient to be sold extensively and to constitute an important article of trade. There was no organized and commercially successful traffic. At the same time the demand for Seneca Oil was limited. It was almost valueless, except as medicine. Some of the settlers along Oil Creek mixed flour with oil to give it body and used it as axle grease. When the supply of lard oil ran out at their sawmill, Brewer, Watson and Company, a large lumbering firm at Titusville, occasionally applied crude oil to the machinery. They found that in using petroleum the journal of the large saw would run "cooler" and that the boxes never "gummed up." Because of the superior lubricating quality, it was not long before they used it exclusively. Those inhabitants unable to secure or afford spermaceti oil or tallow candles burned the oil in crude lamps even though it emitted black smoke and had a disagreeable odor.[31] As early as 1814, De Witt Clinton, president of the Literary and Philosophical Society of New York, in referring to the famous spring near Cuba, New York, suggested to the Society that it might be of considerable consequence to discover whether the petroleum from it might not be used for lighting the cities of the United States.[32] The same idea was warmly advocated in a Pittsburgh paper in 1828. The writer pointed out that the price was low; it could be collected with scarcely any labor; and it was running to waste on Oil Creek.[33] Petroleum would, therefore, be the cheapest, best, and most economical light. Some of it was being used at the time in New York in storage cellars, foundries, and other places where the odor and smoke didn't bother. Both Brewer, Watson and Company and the

[31] *Venango Spectator* (Franklin, Pa.) , May 30, 1855; J. T. Henry, *op. cit.,* 24.

[32] De Witt Clinton, *An Introductory Discourse Delivered Before the Literary and Philosophical Society of New York, on the Fourth of May,* 1814 (N.Y., 1815) , 38.

[33] Quoted in *Oil City* (Pa.) *Register,* April 6, 1865.

Hyde Brothers, another firm engaged in the lumber business along Oil Creek, began lighting their mills with it about 1850. If only an adequate supply could be developed and the offensive odor and smoke eliminated, petroleum had possibilities as a lubricant and an illuminant.

CHAPTER II

Experimenting with New Illuminants

DURING the two decades prior to the American Civil War, scientists were hard at work on the problem of supplying the world with a cheap, safe, and efficient illuminant and lubricant. Whale oil, beeswax, and tallow candles had been used extensively for lighting, and lard oil for greasing purposes; but with the decline of the whale fisheries and an inadequate supply of lard oil, the shortage of oil became increasingly serious in all parts of the world. In the United States, the invention of new machines, the rapid growth of factories, the extension of the railroads, and the inauguration of steamship transportation were making heavier and heavier demands upon the supply. Our trade in oil in the early fifties called for nearly 500,000 barrels of whale oil, 300,000 barrels of lard oil, and about an equal amount of tallow.[1] In the face of an ever greater demand, the price of whale oil increased; it cost about two dollars or two dollars and one-half a gallon in 1850 and many expected it would reach five dollars.[2] Since the price advanced so rapidly that it was almost beyond the average consumer, people were forced to use various burning fluids. Camphene, the most popular and widely used substitute, was a mixture of alcohol and turpentine. The great

[1] Thirty-one and a half gallons to the barrel. James Dodd Henry, *History and Romance of the Petroleum Industry* (London, 1914) , 79.
[2] John Ise, *The United States Oil Policy* (New Haven, 1926) , 8; *Venango Spectator*, May 30, 1855.

disadvantage of all burning fluids, however, was that the use of them involved great risks, and accidents of a serious nature frequently occurred. The need for a new illuminant and lubricant was, therefore, apparent, and between 1840 and 1850 scientists made scores of experiments and spent large sums of money, chiefly in England and Scotland, in trying to find a new source of artificial light.[3]

Foremost among those at work on the problem was James Young of Glasgow.[4] In 1847, a London friend called Young's attention to a petroleum spring in Ridding's colliery at Afferton, Derbyshire, and suggested that he might be able to use the petroleum for some purpose. The spring was then producing about 300 gallons daily. Young, in partnership with his friend Edward Meldrum, purchased the yield, and they proceeded to manufacture illuminating and lubricating oils until 1851. As the supply of petroleum diminished Young experimented and invented a process by which a good illuminant could be produced cheaply by the distillation of coal. Some time previous to this, however, Young found that Torbane Hill mineral, or Boghead coal, gave a better yield of oil than any other coal. Encouraged by the success of his experiments, Young, together with Meldrum and Edward William Binney, formed E. W. Binney and Company at Bathgate and E. Meldrum & Company at Glasgow in the summer of 1850 for the purpose of manufacturing and marketing coal oil. Since there was no satisfactory lamp in which to burn the product without smoking, it was not extensively sold until 1856.

On this side of the Atlantic, Abraham Gesner, a Canadian geologist, was the first to produce an illuminating oil from coal. About 1846 he obtained oil from albertite, a bituminous

---

[3] James Dodd Henry, op. cit., 68.
[4] P. J. Hartog, "James Young," Dictionary of National Biography (Sidney Lee, ed., N.Y., 1909), XXI, 1291–1294; S. F. Peckham, "Production, Technology, and Uses of Petroleum and Its Products," House Misc. Doc., No. 42, 47 Cong., 2 sess., XIII, part 10, 9–10.

mineral from Albert County, New Brunswick, and exhibited samples of it during some public lectures delivered on Prince Edward Island.[5] At first Gesner called his product "keroselain" from the Greek words meaning "wax" and "oil" and suggested by its nature. Later, it was shortened to "kerosene," and the name was extended to all mineral illuminating oils. Gesner came to New York in 1853, took out several patents for the manufacture of kerosene from coal, and later sold the patents and registered title to the New York Kerosene Oil Company, whose large works were erected under his supervision. By 1854 it had commenced the manufacture of kerosene, using cannel coal from England and the United States.

In the meantime, two Boston chemists, Luther and William Atwood, had been doing some research, and in 1852 produced from coal tar the first coal oil made for sale in the United States.[6] When mixed with cheap animal and vegetable oils, it proved to be a fine lubricant for machinery. Since the discovery seemed to be such a coup in chemical science and since the coup d'état of Louis Napoleon occurred only a few months before, the Atwoods called their product "Coup oil." Within a short time, they organized and established the United States Chemical Manufacturing Company at Waltham, Massachusetts, for the purpose of manufacturing "Coup oil," and engaged another chemist, Joshua Merrill, to introduce it to the market.[7]

Attracted by the work of Young, Gesner, and especially the Atwoods, Samuel Downer, a Boston manufacturer of sperm-oil candles, turned his attention to kerosene, and in

[5] James Dodd Henry, loc. cit.; "Abraham Gesner," Appleton's Cyclopaedia of American Biography (James G. Wilson and John Fiske, eds., N.Y., 1888), II, 632–633.

[6] Peckham, op. cit., 9.

[7] Harold U. Faulkner, "Joshua Merrill," Dictionary of American Biography (Allen Johnson and Dumas Malone, eds., N.Y., 1928–1936), XII, 562–563; "Death of Joshua Merrill," Boston Evening Transcript, Jan. 15, 1904.

1854 purchased the United States Chemical Manufacturing Company, which held the patent for "Coup oil." [8] Luther and William Atwood and Joshua Merrill associated themselves with Downer, and they proceeded to make an excellent heavy lubricating oil which, however, had an offensive odor and lacked uniform quality. As a result 90 per cent of the business soon dropped off, a part of the works closed, and ruin faced them.[9]

About this time George Miller and Company of Glasgow appealed to Downer for permission to use the process, manufacture and sell the oil in Great Britain. Downer sent Luther Atwood and Joshua Merrill to superintend the erection of the works and introduce the product. While there, they learned that Miller and Company purchased Bathgate naphtha from James Young's company at sixpence a gallon and mixed it with coal-tar naphtha for making waterproof goods. Somehow or other Atwood conceived the idea of cracking the Bathgate naphtha into a light hydrocarbon suitable for illuminating purposes; he refined five gallons into a water-white illuminating oil, and this, burned in a lamp, produced a light of great brilliance and beauty, without any odor. Learning of Atwood's discovery, Young's company declined to sell any more naphtha and started to manufacture coal oil on a large scale.

Merrill and Atwood had gone to England to help produce lubricating oil, but they returned in 1856 with a determination to manufacture an illuminant. Though angry at their lack of interest in lubricating oil, Downer, nevertheless, followed their advice and embarked heavily upon the manufacture of kerosene for illuminating purposes. Near Boston he erected a most extensive plant, and at Portland he put

---

[8] Faulkner, "Joshua Merrill," *Dic. of Amer. Biog.*, XII, 562; "Samuel Downer," *Dic. of Amer. Biog.*, V, 415; *Boston Evening Transcript*, Sept. 21, 1881.

[9] James Dodd Henry, *op. cit.*, 91.

up another, costing about $500,000 and $250,000 respectively.[10]

Atwood and Merrill carried on ceaseless experiments. Merrill produced a light coal-oil product from Trinidad bitumen, which, on the introduction of the Knapp and Dietz lamps, burned freely in them, producing a bright and beautiful light.[11] He then began using sulphuric acid and alkali as a final process to deodorize and bleach the oil. During 1857 the first attempts were made to use albertite as a source of illuminating and lubricating oil. After much research, Merrill succeeded in perfecting his apparatus for that purpose and erected six retorts, each containing 1,200 pounds of coal and yielding about 360 gallons of crude oil every twenty-four hours. Fifty retorts were soon in operation, producing at the rate of 900,000 gallons of crude oil, or 650,000 gallons of refined oil, each year.[12]

While Downer's kerosene plant constituted the largest and the most important in the United States, others had been established in different parts of the country. Several of the larger ones were located in New York and Brooklyn.[13] They and others along the eastern seaboard used Boghead coal from Scotland and albertite from Nova Scotia. Beginning in 1856, the Breckinridge works on the Ohio River started to manufacture kerosene from cannel coal; and within four years about twenty-five similar plants had been established in Ohio alone, each with a working capacity of 300 gallons a day.[14] The Lucesco plant, in Westmoreland County, Penn-

[10] *Ibid.*, 37.

[11] S. Dana Hayes, "History and Manufacture of Petroleum Products," an address before the Massachusetts Institute of Technology Society of Arts on March 14, 1872. A newspaper account may be found in the records of the Society in the Library at M.I.T.; the whole of it is given in J. T. Henry, *The Early and Later History of Petroleum, etc.,* 186–199.

[12] Hayes, *op. cit.*

[13] *Derrick's Hand-Book*, I, 14.

[14] Kennedy, "Preliminary Report on the Eighth Census, 1860," *House Exec. Doc.* No. 116, 37 Cong., 2 sess., 73. The principal places in Kentucky, Ohio, and West Virginia for manufacturing coal oil from cannel coal were:

sylvania, was the largest cannel-coal works, having a capacity of 6,000 gallons a day.[15] By 1859, between fifty and sixty firms were manufacturing coal oil, representing an investment of nearly $4,000,000, and the product was considered to be equal in every way to that of the British.[16] The manufacture of coal-oil lamps formed the principal business of sixteen companies, which employed 2,150 men, 400 women and boys, and provided work for 125 looms in making lamp wick.[17] As the new industry rapidly developed and assumed larger proportions, kerosene displaced burning fluids of all kinds because it was safer and more economical.

Almost simultaneously with the rise of the coal-oil industry, petroleum was placed on the market as an illuminating material through the efforts of Samuel Kier, an owner and operator of canalboats between Pittsburgh and Philadelphia.[18] For years the salt wells of his father near Tarentum, Pennsylvania, on the Allegheny River, about twenty miles above Pittsburgh, had yielded a small amount of petroleum. About 1846, however, his father drilled a salt well in which oil appeared in a greater quantity. Lewis Peterson, Jr., who managed his father's wells near the Kier wells, experienced the same trouble. In fact, more oil came from Peterson's wells than from Kier's. It was an unwelcome substance, a source of great annoyance, and it is said that Peterson offered a reward to anyone who could utilize it.[19] Without any knowledge as to how it might be used, they allowed it to run off either into the old Pennsylvania Canal or onto the ground.

Canfield and Newark, Ohio; Richie Mines, Peytonia, and Cannelton, West Virginia and Maysville and Cloverport, Kentucky.

[15] Peckham, *op. cit.*, 10.
[16] Kennedy, *loc. cit.*
[17] *Ibid.*
[18] Peckham, *op. cit.*, 10; Kennedy, *op. cit.*, 72; Carl W. Mitman, "Samuel M. Kier," *Dic. of Amer. Biog.*, X, 372; *Warren Mail* (Warren, Pa.), April 1, 1865; *Pittsburgh Dispatch*, Aug. 7, 1892.
[19] James Dodd Henry, *op. cit.*, 142.

One day Peterson took a sample of the oil to the Hope Cotton Factory in Pittsburgh, and, by experimentation, the proprietor found it to be a better and cheaper lubricant for the finest cotton spindles than the best sperm oil. As a result, Peterson agreed to furnish two barrels a week, and for the next few years the factory used the oil, unknown to anyone save the proprietor.

About 1849 Samuel Kier conceived the idea of bottling and selling the oil from his father's wells as medicine. According to the story, his wife had been ill with consumption and her physician prescribed "American Oil," which seemed to help her.[20] Kier compared it with the oil obtained from his father's salt wells, and inasmuch as they possessed the same odor, he concluded that they were identical. This experience and a knowledge of its long usage in western Pennsylvania led Kier to open an establishment in Pittsburgh where the oil was put up in half-pint bottles and wrapped in a descriptive circular telling of the wonderful curative properties of "Kier's Petroleum, or Rock Oil."[21] Through agents who traveled over the country, it was offered to the public. Despite the low price Kier could not dispose of the two or three barrels daily produced by the wells; so he withdrew the salesmen and sold the oil through druggists. With a decline in sales and a supply of oil which exceeded the demand, Kier concluded that something leading toward a more general utilization of the oil must be done.

Having burned the crude oil at the Tarentum wells, Kier believed he might use the surplus if only some method could be found to eliminate the smoke and odor. He sent a man to England to learn the method of refining coal oil, but his agent returned empty-handed, for the British had covered up their operations so well that no idea of the process could be

[20] J. T. Henry, *op. cit.*, 56–58; Peckham, *loc. cit.*
[21] J. T. Henry, *op. cit.*, 56; *Derrick's Hand-Book*, I, 11; *Pittsburgh Dispatch*, Aug. 7, 1892.

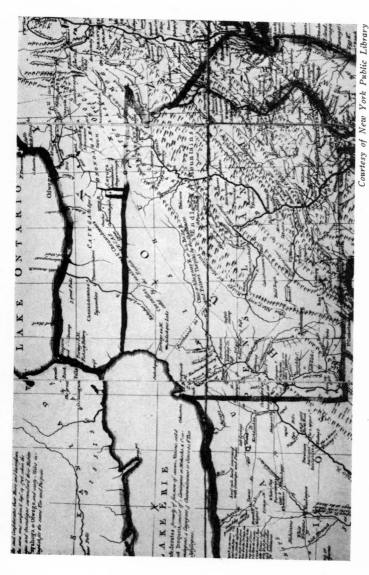

A PORTION OF LEWIS EVANS'S MAP OF 1755

The presence of petroleum is indicated in northwestern Pennsylvania and eastern Ohio.

CIRCULAR USED BY KIER TO ADVERTISE PETROLEUM AS A MEDICINE

obtained.[22] Samples of the oil were then submitted to J. C. Booth, a prominent chemist in Philadelphia, who, after an analysis, recommended that Kier offer the oil to a New York gutta-percha firm, which was seeking a proper solvent for this gum. Kier followed the advice, but the experiments proved unsatisfactory. Upon further reflection, Booth became convinced that by distilling the oil he could obtain an excellent illuminant. He furnished drawings, and Kier immediately erected a refinery containing a one-barrel still, on Seventh Avenue above Grant Street in Pittsburgh. After much experimentation Kier devised a crude distillation process, about 1850 began to distill petroleum, and became America's pioneer oil refiner and industrialist. He named the new product "carbon oil," and sold it for $1.50 a gallon.[23]

Without a suitable lamp in which to burn carbon oil, Kier is reported to have offered $1,000 for a new one that would successfully burn his illuminant; but none appeared.[24] He therefore made some slight changes in the existing camphene lamps, so that they would burn his product without smoking and give off a strong and brilliant light. Since it was better and cheaper than existing illuminants, carbon oil soon came into general use in many places in western Pennsylvania, and Kier had to install a five-barrel still in order to supply the demand.

Afraid of explosion and fire, residents in the vicinity of the refinery complained to the authorities, and the city council gave Kier notice to move his refinery outside the city.[25] Moving to Lawrenceville, a suburb of Pittsburgh, Kier continued his improvements in the quality of the oil and in the adaptability of the lamps. Subsequently he perfected but

[22] James Dodd Henry, *op. cit.*, 105.
[23] The heavier oil secured by distillation was sold to a woolen factory at Cooperstown for cleansing wool.
[24] James Dodd Henry, *op. cit.*, 105.
[25] *Pittsburgh Dispatch*, Aug. 7, 1892; Edwin C. Bell, *History of Petroleum* (Titusville, 1900), 150.

did not patent a four-pronged burner to fit any lamp, and it produced a steady flame with his oil; but the disagreeable odor still remained.[26]

One day in 1857, A. C. Ferris, a New York business man, while visiting the drugstore of Nevin, MacKeown & Company in Pittsburgh, saw in the basement a tin lamp burning carbon oil, which the firm of MacKeown & Finley had been distilling on a small scale from the Tarentum wells.[27] Impressed by its possibilities, Ferris immediately arranged to have a sample tin of five gallons sent to him in New York. On arrival, it was found that the can had leaked, and the straw in which it had been packed in the box was thoroughly saturated with petroleum. It diffused such a penetrating and nauseous smell that the box had to be thrown away. Another shipment consisting of two barrels of oil and some lamps arrived in December, 1857.[28] In order to prevent the oil from penetrating through the wood, the insides of the barrels had been coated with glue, but these precautions proved inadequate. A special storeroom had to be rented at 184 Water Street, and even there the smell became so offensive that a general complaint of the neighbors caused Ferris to move to 192 Pearl Street. Ten more barrels soon arrived, making a total of twelve barrels received during 1857.[29]

Ferris now arranged to secure a larger amount of oil; MacKeown & Finley agreed to supply him with about 2,000 barrels of oil a year.[30] His attention having been called to the fact that Kier distilled petroleum, Ferris also purchased and received a shipment from him in March, 1858. Upon testing it, Ferris wrote that "it is not as good and saleable a

---

[26] *Pittsburgh Dispatch*, Aug. 7, 1892.

[27] James Dodd Henry, *op. cit.*, 105.

[28] *Ibid.*, 75. The barrels contained eighty-four and a half gallons each. For the invoice on this shipment see *Ibid.*, 107.

[29] Five barrels containing 214 gallons arrived on Dec. 18, 1857, and five more on Dec. 23. *Ibid.*, 75.

[30] The agreement was made on Feb. 11, 1858, but later it was extended for two years beginning Jan. 1, 1859. James Dodd Henry, *op. cit.*, 107.

colour as MacKeown and Finley's, being a deep blue." [31]
Furthermore, it had a very obnoxious odor. "In almost every
instance where we have sold it," Ferris wrote to Kier, "we
have had complaint of it, and a good deal of it returned upon
our hands with expenses." [32] Since Ferris had not noticed the
odor in Pittsburgh, he concluded that when the oil was run
from the still directly into the barrels, the gases, instead of
passing off, remained and affected the oil so as to render it
exceedingly offensive. In order to improve the smell, Ferris
advised Kier to run it into his cistern, allow it to cool, and
expose it to the air. Ferris bought one hundred additional
barrels from Kier during the spring, but the oil continued to
be so unsatisfactory that no further purchases were made. [33]

At first Ferris sold carbon oil in New York just as it came
from the stills in Pittsburgh without any further treatment.
Canvassers, with lamps in one hand and a can of oil in the
other, entered different places of business, exhibited the light
to the proprietor and visitors, then sought permission to send
a few gallons and a half-dozen lamps on trial. The proprietor
had the privilege of returning everything or paying the bill.
Often those who purchased carbon oil made blunders in
managing the lamps, and then they called upon Ferris to
indemnify them for soiled or damaged furniture. Nevertheless, his advertisements about the new oil aroused so much
public curiosity that he was overwhelmed by visitors and had
to employ a young man to receive guests and answer questions.

Since carbon oil from Pittsburgh would not burn very
well in the lamps, Ferris proceeded, with considerable effort
and much expense, to adapt it to "a lamp in which it could
be burned with public approval." Finding a burner that

[31] *Ibid.*, 110.

[32] *Ibid.*

[33] Later, Charles B. Holmes of New York, who had acquired a knowledge of refining coal oil while with the Portland Kerosene Company, went
to Pittsburgh at the suggestion of Ferris, and instructed Kier in the art of
refining oil. James Dodd Henry, *op. cit.*, 110–11.

would operate with some degree of success, he discovered that it was incapable of consuming more than half the oil in the lamp, when the flame would drop, the light almost cease, and much smoke and smell penetrate the room. This was partially overcome by the admixture of distilled resin oil, the use of which he continued many months; but in the end, by improving and refining the oil and perfecting the burner, he overcame this difficulty. The next step was to deodorize the oil, which he did by applying a very hot and strong solution of caustic soda. Improving the odor satisfied his customers for a short time, but the color, an orange or dark straw, caused many complaints; so Ferris applied sulphuric acid and then caustic soda, which turned it to a light lemon color.

With an increasing number of customers and the spread of knowledge about petroleum, Ferris retired in May, 1858, from the business of the Hudson Mills, with which he had been connected for some time, in order to give his sole attention to marketing petroleum, improving its quality, and perfecting the lamps. The trade in petroleum in 1858 involved about 1,183 barrels, almost all of which was handled by Ferris.[34] In fact, his success in introducing carbon oil in New York, cultivating the market, and producing an improved quality of illuminating oil from petroleum outran the sources of supply.

The price jumped from seventy-five cents a gallon to $1.50 and then to $2.00, a price so high that it furnished an incentive to find petroleum, if possible, in greater quantities. Charles Lockhart of Pittsburgh and a friend in Tarentum formed a partnership in 1853, bought a salt well, across the river from Tarentum, which was producing petroleum, and sold the oil to Kier. Lewis Peterson, Jr., likewise concluded that he would improve the opportunity for getting oil. He

[34] *Derrick's Hand-Book*, I, 16. Two wells at Tarentum furnished 1,008 barrels and the rest came from Canada; from 1857 to 1860 Ferris claimed that he handled 30,000 barrels of oil a year. James Dodd Henry, *op. cit.*, 105.

FRANCIS B. BREWER

GEORGE H. BISSELL

and a friend decided to buy a salt well on the Humes farm, not far from the Petersons' property. Believing that if they enlarged the well to the size of others in the vicinity it too would pump oil, they purchased the whole farm for $20,000 and paid $20,000 for the lease on which the well stood. After they had reamed out and enlarged the hole, the well yielded two and one-half to five barrels a day. The entire product was sent to Baltimore for use at the carding mills. The income of Peterson and his friend from this well for 1858 was said to be about $10,000. In order to increase production, they decided to sink a shaft to the source of the well, possibly 300 or 400 feet below the surface. They secured a steam engine, the necessary apparatus, and experienced drillers, and commenced operations. The necessity of constantly pumping out the water, the danger from fire in the steam boiler, and the lack of any favorable sign led them to abandon the well at about 200 feet. They sold the tract to the Tarentum Oil, Coal and Salt Company, which continued the work. Another party drilled a well at Greensburg, Pennsylvania, in March, 1858, to a depth of 400 feet and then abandoned the hole. All efforts to increase the supply of petroleum, however, met with indifferent success until a country doctor started a chain of events which led to the drilling of the Drake well.

CHAPTER III

Launching the First Petroleum Company

THE birth of the petroleum industry took place in the northwest corner of Pennsylvania, where the mountain ranges of the Alleghenies begin to flatten out toward the central plains. Running from north to south through these foothills and eventually flowing into the Allegheny River was Oil Creek; a short distance above the point where Pine Creek flowed into Oil Creek was the quiet village of Titusville, with 300 to 400 inhabitants. To this place Francis Beattie Brewer, a graduate of Dartmouth College and a practicing physician, moved in 1851 from Massachusetts to join the firm of Brewer, Watson and Company. Immediately he became interested in the old oil spring located near Upper Mill on the company's property, about two miles below Titusville and within a few rods of Oil Creek. A thorough examination of the spring and other surface conditions along the entire length of Oil Creek convinced Dr. Brewer that the oil had great possibilities, and after much discussion with other members of the firm he persuaded them that they should utilize the oil and make it profitable.[1] As a result, Brewer, Watson and Company made a lease with J. D. Angier of Titusville on July 4, 1853, the first lease in the United States in connection with the development of the petroleum business.[2] Angier agreed for five years to repair and keep the spring in order, construct new springs, and gather oil; when the expenses had

[1] *Titusville Morning Herald*, Jan. 28, 1881.
[2] J. T. Henry, *The Early and Later History of Petroleum, etc.*, 60–61.

been deducted from the proceeds, the balance, if any, was to be equally divided. Angier dug rude trenches to convey the oil and water to a central basin, and by means of some crude and simple machinery, erected at a cost of about $200, separated the oil from the water and thus increased the supply of petroleum for lighting the mills and lubricating the machinery. Three or four gallons a day and sometimes as many as six were collected.

In the fall of 1853, Dr. Brewer carried a small bottle of petroleum on a trip to Hanover, New Hampshire, to visit relatives and friends. While he was there, both Dr. Dixi Crosby of the Dartmouth Medical School and Professor O. P. Hubbard of the Chemistry Department of Dartmouth College examined the sample.[3] The latter at once pronounced it very valuable but, because the oil could not be obtained in large quantities, said that it could hardly become an article of commerce. A few weeks later, George H. Bissell, another graduate of Dartmouth and a young lawyer in New York, returned to his home in Hanover, saw the bottle of petroleum in Crosby's office, and immediately became interested in its possibilities.[4] In view of the similarity between petroleum and coal oil, Bissell wondered why the former could not be used as an illuminant, provided a sufficient supply could be found. Excited over the possibilities of a large speculative enterprise, a permanent business, and a certain fortune, Albert H. Crosby, son of Dr. Crosby, induced Bissell to pay his expenses to Titusville in the summer of 1854 to inspect the oil springs and land.[5] If he returned with a favorable report,

[3] *Titusville Morning Herald*, Jan. 28, 1881. The famous bottle of rock oil remained a long time in the Dartmouth Medical School building, but it was later transferred to the Chemistry Department, which lost it.

[4] Frank A. Taylor, "George Henry Bissell," *Dictionary of American Biography* (Allen Johnson and Dumas Malone, eds., N.Y., 1928–1936), II, 301–302; J. T. Henry, *op. cit.*, 61; S. S. Hayes, "Report of the United States Revenue Commission on Petroleum as a Source of National Revenue," *House Executive Document* No. 51, 39 Cong., 1 sess., 4.

[5] *Titusville Morning Herald*, Jan. 28, 1881.

Bissell and his business partner, Jonathan G. Eveleth, agreed to organize a company, buy the land, develop the springs, and market petroleum.[6]

About thirty days later, Crosby visited Titusville and, in company with Dr. Brewer, examined the oil springs.[7] They traveled as far down Oil Creek as Hamilton McClintock's farm to see the most famous of all the early oil springs. "As we stood on the circle of rough logs, surrounding the spring," Dr. Brewer wrote, "and saw the oil bubbling up, and spreading its bright and golden colors over the surface, Crosby at once proposed to purchase the whole farm, which we could have done for $7,000, but not enough money. When I told Crosby that we [Brewer, Watson and Company] did not want to take money from the lumber business to put into oil, Crosby said, 'damn lumber, I would rather have McClintock's farm than all the timber in Western Pennsylvania.' "[8] On their return to Titusville, Brewer, Watson and Company, through Dr. Brewer, agreed to sell the Hibbard farm of about one hundred acres, embracing the principal spring, for $5,000, if Bissell and Eveleth would organize a company to develop the property.[9] In addition, oil rights in the surrounding lands of Brewer, Watson and Company consisting of 1,100 or 1,200 acres would be included. The joint stock company should have a capital of $250,000: one-fifth to go to Brewer, Watson and Company; one-fifth to be treasury stock; and three-fifths to be divided as thought best by the original purchasers.

With several samples of petroleum, Crosby enthusiastically and hopefully left Titusville for New York, where he met

[6] Eveleth and Bissell were at this time promoting the sale of stock in the American & Foreign Iron Pavement Company and the Safety Railway Switch Company.

[7] The details about Crosby's visit to Titusville are given in a short manuscript written by Dr. Brewer in the Brewer Papers. Hereinafter this work will be referred to as Brewer's Account.

[8] *Titusville Morning Herald*, Jan. 28, 1881.

[9] *Ibid.* See also Brewer's Account.

Bissell, made an exceedingly favorable report, and explained Dr. Brewer's proposal. Within a few days Crosby sent a telegram, and later a letter, informing Dr. Brewer of Eveleth and Bissell's acceptance. Before departing for Hanover, Crosby saw one or two brokers, and, as a result, his confidence in the ultimate success of the venture was considerably increased. "I do not believe," he confidentially wrote Dr. Brewer, "that there will be any necessity of rushing the thing through at 40 horse power and making it pay all that it can in three months, as it *must* increase in value as it is better known and by securing the influence of the Press in giving details of its history and the manifold uses it can be put to we shall be able to bring the stock actually above par in one year's time." [10] He asked Dr. Brewer to send some oil to New York for exhibition purposes; and by the time it arrived Crosby expected that they would have circulars, stock books, and everything else ready to issue to "a gullable public."

Angier immediately gathered and shipped three barrels of oil to Eveleth and Bissell, whose office was located at the corner of Broadway and Franklin Street in the magnificent new building of D. Appleton and Company, the leading publishers and booksellers of the city.[11] In the absence of Eveleth and Bissell the drayman left the barrels in front of this salesroom. The store was a model of elegance and taste, filled with rare books and literary treasures, and ornamented with costly works of art from every land. Surprised and annoyed were the fastidious proprietors, to find these disgusting looking and vile smelling barrels on the pavement in front of their store. Standing in the hot sun, oil oozed from every pore of the barrels, sending off small streams along the stone pavement. It was a most unattractive sight. Appletons had the barrels loaded into a passing dray and hauled to an un-

---

[10] Crosby to Dr. Brewer, Sept. 11, 1854, in Brewer Papers. See also Brewer's Account.

[11] *Titusville Morning Herald*, Jan. 28, 1881.

known destination. So tenacious were the stones of their hold on this fluid and so offensive was the smell to the aesthetic taste of the proprietors of the building that even some of the stones were removed. When Eveleth and Bissell returned and learned what had happened, they made a diligent search for the oil. After several weeks of effort, they located the barrels and took a portion of the oil to their office. An unsightly spot soon appeared on the white ceiling of the store below, and it smelt to high heaven. But over this spot, in that office, in that building, arrangements were soon made to launch the first petroleum company.

Late in September, everything was in readiness to put the stock on the market, except the signing of the contract, and Dr. Brewer, empowered as attorney for Brewer, Watson and Company, departed for New York City to complete the transaction. In the first interview, Eveleth and Bissell "candidly confessed that they could not believe the statements made by Crosby relative to the existence and abundance of the oil." [12] After learning what Crosby had reported, Brewer gave them the facts in greater detail—a description of the country, the indications of oil, its abundance, and the manner of gathering it. Despite Brewer's convincing statements, the whole thing looked too visionary, and Eveleth and Bissell refused to go ahead with the original plans even though they had incurred obligations to the extent of several hundred dollars for seals, stock certificates, stock books, and other items.[13] Under the circumstances, Dr. Brewer left the office thinking that the matter had been settled and decided to get in touch with some Pittsburgh parties, who had been previously interested in obtaining the oil. Before Brewer left the city, however, Eveleth and Bissell sent a note to his hotel asking him to call. As a result of this subsequent meeting, Dr. Brewer made them an offer: his proposition should

[12] Brewer's Account; J. T. Henry, *op, cit.*, 65.
[13] Brewer's Account.

stand open, and no money should be paid, till one of the firm came to Titusville for a personal examination of the property.[14] If they found Brewer's statements true the bargain would be consummated, and if not Brewer, Watson and Company would pay all expenses of the trip. To this, Eveleth and Bissell readily assented.

Before leaving for Titusville, Eveleth and Bissell went to New Haven, Connecticut, at the request of Anson Sheldon, a retired minister there, to interest certain capitalists in buying the oil tract in Titusville.[15] Stopping at the Tontine Hotel, the local gathering place for all who sought the latest news, they met James M. Townsend, president of the City Savings Bank, and other prominent citizens. Ambitious, energetic, and eager to try something new and exciting, Townsend listened with unusual interest to Bissell's stories about the oil springs at Titusville and the possibilities of petroleum's supplanting coal oil, provided capital could be found to develop the land. While discussing the matter with Townsend and others, the proprietor of the Tontine learned that Eveleth and Bissell could not pay their bill and almost turned them out; but Townsend begged him not to do so as they seemed to be gentlemen, though poor, and he arranged to pay the bill if they did not meet the obligation.[16] In the end, Bissell fired the enthusiasm of Townsend and others in New Haven, who saw the possibilities of making money out of petroleum. Prior to making any investment, however, the New Haven men thought they should do two things: send a committee to Titusville to see the land, and have a scientific analysis of the oil made in order to determine its economic

---

[14] *Ibid.*

[15] James M. Townsend, "A Brief Account of the Development of the Petroleum Industry in Pennsylvania," a manuscript among the Townsend Papers, Drake Museum, Titusville. Also see Henry M. Townshend, *New Haven and the First Oil Well* (New Haven, 1934).

[16] *Ibid.* James M. Townsend's feelings were considerably injured when Bissell later became wealthy and failed to recognize his New Haven acquaintances on the streets of New York.

value. Eveleth and Bissell, therefore, arranged to have two eminent chemists, Luther Atwood of Boston and Professor Benjamin Silliman, Jr., of Yale College, analyze the oil.[17] Although the cost might be very heavy, the analysis, if favorable, would stimulate the sale of stock. Atwood reported within a short time on the excellent quality of the oil and indicated the uses to which it might be applied.[18] Since Professor Silliman gave it a longer and more thorough analysis, his report was delayed.

Eveleth now proceeded to Titusville in order to determine whether or not the quantity of petroleum seemed sufficient for commercial exploitation.[19] Driven by R. D. Fletcher, a clerk of Brewer, Watson and Company, to the site of the oil spring at Upper Mill, they dug up considerable dirt and threw it into the water at the tailrace and watched the oil float off upon the surface. Eveleth then carefully examined all the territory included within the proposed contract. Pleased with the results of his investigation and entirely satisfied about the abundance of oil, Eveleth at once predicted a large flourishing business. In fact, he was fully as enthusiastic as Crosby had been and returned to New York to arrange the final details.

On November 10, 1854, Dr. Brewer deeded the Hibbard farm to Eveleth and Bissell for $5,000.[20] Since the actual purchase price seemed to be such an insignificant fraction of the contemplated capital of the new company that it might handicap the sale of stock, the land was put in at $25,000. Eveleth and Bissell gave their joint and individual notes for the land,

---

[17] Eveleth and Bissell to Dr. Brewer, Nov. 4, 1854. Brewer Papers.

[18] Eveleth and Bissell to Dr. Brewer, Nov. 6, 1854. Brewer Papers.

[19] Brewer's Account; Bissell's account of the visit is given in Hayes, *loc. cit.* Bissell in 1866 said, "We did not prospect the oil for medicinal purposes, but we believed it would be a good illuminator, and we sought it is an article of commerce." R. D. Fletcher's account is given in James Dodd Henry, *History and Romance of the Petroleum Industry*, 152.

[20] The original agreement made on Nov. 10, 1854, is among the Townsend Papers.

JAMES M. TOWNSEND

BENJAMIN SILLIMAN, JR.

and so no cash changed hands.[21] Since Crosby could not meet his share of the expense, he dropped out of the affair. A short time later, Eveleth and Bissell organized the Pennsylvania Rock Oil Company of New York. The certificate of incorporation, filed at Albany on December 30, 1854, showed a capitalization of $250,000, divided into 10,000 shares.[22] The management of the company was placed in the hands of seven trustees, all of whom except Dr. Brewer were mere figureheads, occupying positions it was necessary for appearance's sake that some one should fill. Not more than one of them at most represented stock for which he had paid. As the last step in organization, Eveleth and Bissell conveyed the Titusville property to the trustees on January 16, 1855; but the deed was not recorded, and the land continued in their possession until fall.

The New York promoters now made strenuous efforts to sell stock at almost any price.[23] Much of the burden fell upon Eveleth, for Bissell received an injury late in January, which made it impossible for him to be very active.[24] They gave Crosby, now a New York City newspaper reporter, a few shares for his former services, a few more for using his influence through the press, and 200 shares to sell.[25] He offered his own stock for as low as fifty cents a share to a gentleman in Connecticut, the same person to whom Eveleth and Bissell were trying to sell stock at $2.50 a share. Having spent $7,000 to $8,000 without any return, and calculating on this sale to help out on obligations, Eveleth and Bissell were thrown into consternation and immediately bought all of Crosby's stock,

---

[21] "I say papers were executed," wrote Dr. Brewer, "for it was all on paper and no money was paid down but notes were given." Brewer's Account.

[22] J. T. Henry, *op. cit.*, 70, gives the certificate of incorporation in full; the *Venango Spectator*, May 30, 1855, contains a short article about the formation of the company.

[23] Speaking about the activities of the promoters, Dr. Brewer wrote: "The financial dodges which were enacted during the next few months to get things in motion would do credit to the cleverest Wall St. operator in the arena where Bulls & Bears strive for mastery." Brewer's Account.

[24] Eveleth to Dr. Brewer, Feb. 1, 1855, and Feb. 8, 1855. Brewer Papers.

[25] J. T. Henry, *op. cit.*, 72–73.

eliminating him from the venture.[26] They then secured the services of Anson Sheldon to fan the flame of interest in New Haven. Sheldon became enthusiastic, even fanatical, about the future success of the company. He bought several hundred shares, for which he gave his note, and received 2,000 shares to sell at a minimum fixed price.

Despite herculean efforts, the stock did not sell. Owing to hard times, it was difficult to get men to invest. "Money was never in more demand and business men are sorely pressed," Bissell wrote to Dr. Brewer.[27] To make it more difficult, there was an almost universal distrust of the enterprise. In Titusville, they dubbed the petroleum company "The Fancy Stock Company." Paying $5,000 for a piece of land that was hardly worth the taxes upon it seemed preposterous to the local citizens, and few people were willing to invest a dollar, though many were abundantly able.[28] Elsewhere investors refused to invest because they either knew little or nothing about petroleum, or were not sure of its economic value, or were not certain about the quantity in which it could be found. Equally important, if not more so, was the total lack of confidence in the financial responsibility of those managing the company. Indicative of the general feeling toward Eveleth and Bissell was the statement of Ebenezer Brewer, father of Dr. Francis B. Brewer, who wrote to his son: "I always told you that I had no confidence in the men, from the very nature of the transaction, and all that you would ever get would be what you received in the sale. . . . Now mark well what I tell you, it is for your interest alone that I now say it—You are associated with a set of sharpers, and if they have not already ruined you, they will do so if you are foolish enough to let them do it."[29] Of Eveleth and Bissell, Charles Richmond, Jr., a New York merchant, said, "From what little I

26 Eveleth to Dr. Brewer, Feb. 8, 1855. Brewer Papers.
27 Eveleth and Bissell to Dr. Brewer, Dec. 1854. Brewer Papers.
28 Gale, *The Wonder of the Nineteenth Century*, 67.
29 Ebenezer Brewer to Dr. Brewer, March 23, 1855. Brewer Papers.

have seen of them and their transactions, I have no confidence in them as business men and would not trust them without security." [30] On the other hand, Richmond believed "that if a company could be formed on the right principles by men of Capt [capital] & influence money could be made out of it." [31]

Confronted by all these difficulties, the promoters anxiously awaited Professor Silliman's report and hoped it would spur the sale of stock. Late in December, Professor Silliman gave them an insight into his experiments and some encouragement about the outcome. He wrote: "I can promise you that the result will meet your expectations of the value of this material for many most useful purposes. The oils which I have so far obtained are perfectly fluid when exposed in this coldest weather, are not acid, & do not seem to evaporate, or suffer change by exposure to air. I am having them tried on Watches & fine machinery." [32] His tests soon indicated, however, that the value of the oil depended mostly upon its properties as an illuminant. [33]

Owing to an explosion Eveleth and Bissell had to buy all new apparatus for Silliman, and this delayed the completion of the analysis until April 16, 1855. [34] Since Eveleth and Bissell failed to place $100 to his credit in the Bank of New York, Silliman left his report in the hands of a friend in New York City with instructions not to deliver it until satisfactory arrangements had been made for paying the bill, which amounted to $526.08. [35] Strange to say, here was the first im-

[30] Charles Richmond, Jr., to Dr. Brewer, May 22, 1855. Brewer Papers.

[31] Charles Richmond, Jr., to Dr. Brewer, May 3, 1855. Brewer Papers.

[32] Extract from Professor Silliman's letter dated Dec. 21, 1854. Brewer Papers.

[33] Sheldon to Dr. Brewer, April 10, 1855. Brewer Papers.

[34] The progress of Silliman's experiments can be followed from the letters of Eveleth to Dr. Brewer, Feb. 8 and Feb. 17, 1855, and in that of Sheldon to Dr. Brewer, April 10, 1855. Brewer Papers.

[35] Sheldon to Dr. Brewer, April 23, 1855. Brewer Papers. The bill included the expense for a portion of the apparatus employed in the analysis. About $75 had been previously paid in New York for the retort, lamps, fixtures, ores, etc., making the whole expense of the analysis exceed $600.

portant scientific analysis of rock oil which proved to be a decisive factor in the establishment of the petroleum industry, withheld until arrangements could be made to pay $526.08! As soon as he learned of the predicament, Sheldon left for New York to see if he could not raise the money. He called on numerous friends, all of whom refused to aid.[86] Finally, Eveleth advanced the money out of his own pocket and they secured the long-looked-for report.[87]

Silliman's analysis referred to occurrences of oily fluids coming from the earth's surface in Persia, Baku, the Duchy of Parma, of its hardening into bitumen, asphalt, and mineral pitch, its production in large quantities in India, and its appearance at many points along the Ohio River in the United States. But the Titusville oil, unlike that from any other place, did not become hard and resinous from continued exposure to air. With the meticulous precision of a professor of general and applied chemistry, he then gave the details of his experiments and found that for illuminating purposes, it gave a most perfect flame with the Argand burner. In concluding, Silliman said: "It appears to me that there is much ground for encouragement in the belief that your Company have in their possession a raw material from which, by simple and not expensive process, they may manufacture very valuable products. It is worthy of note that my experiments prove that nearly the whole of the raw product may be manufactured without waste, and this solely by a well directed process which is in practice one of the most simple of all chemical processes."[38] The promoters sent the analysis to a New Haven printer on April 23, and within a few days it was ready for distribution among prospective investors.

The report proved to be a turning point in the establish-

---

[86] Charles Richmond, Jr., to Dr. Brewer, April 20, 1855, and Sheldon to Dr. Brewer, April 23, 1855. Brewer Papers.

[87] Eveleth to Dr. Brewer, April 21, 1855. Eveleth paid Silliman $300 more in June, 1855 (Eveleth to Dr. Brewer, June 17, 1855). Brewer Papers.

[38] J. T. Henry, op. cit., 53; Venango Spectator, May 30, 1855.

EDWIN L. DRAKE

STOCK CERTIFICATE OF THE SENECA OIL COMPANY

Townsend Papers

ment of the petroleum industry, for it dispelled many doubts about its value and caused New York and New Haven capitalists to look with greater favor upon the enterprise. Charles Mulock, a New York business man, was especially pleased with Silliman's report and planned to make a personal inspection of the oil springs at Titusville. If he found that they would furnish an adequate supply under proper management, then he and several others, including his father-in-law and Charles Richmond, Jr., proposed to take all the stock and pay cash.[39] Before buying stock, however, they insisted on a reorganization of the company so that men of means and influence would be in control.

Sheldon pointed out not only the beneficial effect of Silliman's report in New Haven but, at the same time, the greatest obstacle to the sale of stock. "Silliman's Report is now generally in the hands of the monied men in this city," he wrote to Dr. Brewer on May 11, 1855, "& the impression it has made is decidedly favorable to the interest of the Pennsylvania Rock Oil Company, but with the present state of feeling existing here, in reference to joint stock companies formed under the Laws of the State of New York, & doing their business in the city of New York, I do not believe that any amount of stock could be taken by capitalists in this city." [40] In New Haven and elsewhere the history of the New York & New Haven Railroad Company, also the Western Empire Company, was remembered with sorrow; many had been ruined through frauds committed by those companies; others had experienced losses; and it was not strange, therefore, that wealthy men were extremely cautious. If Eveleth and Bissell would reorganize the company in Connecticut, where the property of a stockholder was not liable for the debts of a company as in New York, some of the most prominent New Haven men assured Sheldon that all the stock

[39] Charles Richmond, Jr., to Dr. Brewer, May 3, 1855. Brewer Papers.
[40] Sheldon to Dr. Brewer, May 11, 1855. Brewer Papers.

could be disposed of in that city. Having spent an enormous amount of time in trying to sell stock and having exhausted his means, Sheldon pathetically wrote, "Unless some course is adopted very soon by which the wheels shall be set in motion, to some good effect, I shall be compelled to look out for my bread and butter in some other quarter." [41] In view of the general situation Ebenezer Brewer declared the whole enterprise to be a "perfect failure" and the stock not "worth a straw." [42]

Under the circumstances Eveleth and Bissell decided to abandon the Pennsylvania Rock Oil Company of New York and form a new company under Connecticut law. Almost immediately the New Haven capitalists began to subscribe stock, and by June 25, 1855, two-thirds of the stock had been taken. [43] Sheldon had two offers for the rest, but did not entertain one of the propositions as the party was objectionable. "There is not so much difficulty in selling the stock," he said, "as there is in securing the right kind of men—We need men of means —of good banking habits, and that will work harmoniously together; & this is the class of men that are coming in . . ." [44] Two of the best oil dealers and manufacturers took stock, and it seemed quite likely that an oil firm in Boston would subscribe. In New Haven no two better men than William A. Ives and Asahel Pierpont could have been drawn into the enterprise, for they had a reputation of making a success of everything in which they were interested. Even Professor Silliman agreed to take stock, and the name of this great scientist gave prestige to the enterprise. In fact, all the New Haven subscribers were men of wealth and ranked high in business talents.

When the new company would be organized depended

[41] *Ibid.*

[42] Ebenezer Brewer to Dr. Brewer, June 4, 1855. Brewer Papers.

[43] Eveleth to Dr. Brewer, June 25, 1855, and Sheldon to Dr. Brewer, July 3, 1855. Brewer Papers.

[44] Sheldon to Dr. Brewer, July 3, 1855. Brewer Papers.

largely upon the impressions gained at Titusville by a committee composed of Anson Sheldon and Asahel Pierpont, representing the New Haven group. Early in July, Sheldon and Pierpont arrived in Titusville to inspect the springs. "Do all in your power," Eveleth wrote to Brewer, "to make them appreciate the true value of the springs. Work that *one*, if possible, won't you?" [45] Dr. Brewer hospitably received them, took them to the springs, and both men were exceedingly pleased with the results of their trip. Since Pierpont had other places to visit before he returned home, he told Sheldon to say to his friends back in New Haven that "the oil is there & he was satisfied." [46]

The committee's report had a valuable effect in New Haven: it made subscribers anxious to have the company organized as soon as possible; it increased the number of requests for Silliman's report and samples of oil; it created a strong conviction that the oil would pay a good dividend; and it led many persons to want stock whom Sheldon had not thought of as being interested in the enterprise. The time was ripe, and so they incorporated the Pennsylvania Rock Oil Company of Connecticut on September 18, 1855, with a capitalization of $300,000, divided into 12,000 shares. [47] Eveleth and Bissell retained a controlling share of the stock, but the new company was predominantly a New Haven enterprise: Professor Silliman was president; New Haven was the headquarters of the company; and the by-laws provided that a majority of the board of directors should be chosen from among the New Haven stockholders.

Simultaneously with the launching of the new company, Eveleth and Bissell began closing up the business affairs of

---

[45] Eveleth to Dr. Brewer, June 25, 1855. Brewer Papers.

[46] Sheldon to Dr. Brewer, Aug. 7, 1855. Brewer Papers.

[47] The Articles of Association may be found among the Townsend Papers. On the published list of original stockholders appeared the names of a number of individuals who never took stock, and so it was retained by Eveleth and Bissell.

the old company. On account of the death of Bissell's mother and his subsequent illness, most of the work fell upon Eveleth, who had always given cheerfully and generously of his time and money to make the enterprise a success.[48] He finished paying Silliman for his report, settled all other bills, drew the legal papers, and issued a notice for the stockholders to meet on August 28, for the purpose of transacting whatever business might be necessary and to empower the trustees to sell the land.

Since Eveleth and Bissell had never deeded the land to the Pennsylvania Rock Oil Company of New York, the company did not own any property. Furthermore, the directors stood liable to pay par value for all outstanding stock that had been sold.[49] If they could get in all of the stock, Eveleth and Bissell could then sign over the deed to the new company. Otherwise, all the papers must first be made out to the old company and then transferred to the new one. As most of the stock was held by Eveleth and Bissell or their agents, it seemed easier and wiser to call in all the old stock, which they did. Plans for selling the land were quickly approved, and arrangements made to transfer the property. Inasmuch as the land of any foreign corporation, organized beyond its borders, was liable to be forfeited to the State of Pennsylvania, Eveleth and Bissell deeded the Hibbard farm on October 5, 1855, to Asahel Pierpont and William A. Ives, who gave bond for the value of the property and promptly leased it for a "valuable consideration" to the Pennsylvania Rock Oil Company of Connecticut for ninety-nine years.[50] Within a short time all other arrangements for closing out the old company had

[48] Eveleth to Dr. Brewer, June 17, 1855. Brewer Papers. Eveleth says that he paid Silliman $300 in June. Another letter of Aug. 10, 1855, says that he had already advanced the $1,000 on behalf of the company, while no one else had paid one cent. "It is not using me right," he declared.

[49] Eveleth to Dr. Brewer, Aug. 10, 1855, and Eveleth to Dr. Brewer, Aug. 20, 1855. Brewer Papers.

[50] Among the Townsend Papers is the deed given by Eveleth and Bissell to Ives and Pierpont.

been completed. After almost a year of hard work and worry, it was a joyful occasion for more than one reason: Bissell got married; Eveleth expected an heir in a few days; but, most important of all, the scheme for developing and introducing petroleum to the market had been saved.[51]

[51] Sheldon to Dr. Brewer, Oct. 19, 1855 (Brewer Papers). A meeting of the stockholders of the Pennsylvania Rock Oil Company of New York was held at New Haven on Nov. 23, 1855, for the final adjustment of affairs. Eveleth and Bissell to Dr. Brewer, Nov. 15, 1855 (Brewer Papers). The Company was finally dissolved by a proclamation of the Governor of New York on April 2, 1924.

CHAPTER IV

The Seneca Oil Company and the Drake Well

THE New Haven men took an active interest in the new company and the prospects for a successful venture seemed bright. The publication of the Articles of Association created a favorable impression in New Haven, and an increasing number of citizens believed that there was something valuable on Oil Creek.[1] Many now wanted to buy stock who, a few weeks previously, had pronounced the whole thing a humbug. Townsend and his associates took additional stock, raised a small sum of money for the treasury, and sent Asahel Pierpont, an excellent mechanic, to Titusville to examine the springs with a view to improving upon Angier's method of collecting oil. Pierpont recommended more active operations; but, owing to a lack of harmony which unexpectedly developed between the New Haven and New York stockholders over various things, no more money could be raised for the treasury, so that the company dispensed with Angier's services, abandoned all work, and the auspicious beginning quickly came to an end.[2]

While company affairs were at a standstill, Bissell accidentally stopped one hot summer day beneath the awning of a Broadway drugstore in New York. Attracted by the advertisement beside a bottle of Kier's rock oil in the window,

[1] Sheldon to Dr. Brewer, Sept. 25, 1855. Brewer Papers.
[2] J. T. Henry, *The Early and Later History of Petroleum, etc.*, 80.

Bissell stepped inside, examined the circular, studied the picture of the salt-well derricks, and in this way got an idea that a greater supply of oil might be obtained at Titusville by drilling a well.[3] Puzzled over whether he and Eveleth should undertake the work or induce the company to help, Bissell consulted with Rensselaer H. Havens of Lyman & Havens, a prominent Wall Street real estate broker. Desirous of executing the idea himself, Havens offered Eveleth and Bissell $500 to secure a lease for him from the Pennsylvania Rock Oil Company of Connecticut. Eveleth and Bissell thereupon proposed to the New Haven people that if they would purchase from them an amount of stock equal to one-fourth the stock held in New Haven, then Eveleth and Bissell would find parties to lease and work the land.[4] The New Haven stockholders, after some consideration, agreed to the proposition, purchased the stock, and then subleased the Titusville property to Lyman & Havens for fifteen years for a royalty of twelve cents a gallon on all oil raised.[5]

Before Lyman & Havens had time to commence operations, the panic of 1857 overwhelmed them, so that it was impossible to fulfill the terms of their contract. Seeking a way to be released from its obligations, they discovered that the wives of the grantors from whom the lands had been purchased had failed to sign the papers, and, in case of the death of their husbands, they would be entitled to a dower. Taking advantage of this omission, Lyman & Havens surrendered their lease on the ground of a defective title. Under the circumstances, the New Haven people were inclined to believe, and

---

[3] J. T. Henry, *op. cit.*, 83. Professor Silliman frankly confessed that during the five months of his experimentation the thought of boring for oil as for water had not occurred to him. *Ibid.*, 83; see Bissell's story in Hayes, "Report of the United States Revenue Commission on Petroleum as a Source of National Revenue," *House Exec. Doc.* No. 51, 39 Cong., 1 sess., 4–5.
[4] Townsend to Brewer, Watson and Company, Jan. 8, 1858. Townsend Papers.
[5] Lease of the Pennsylvania Rock Oil Company of Connecticut to David H. Lyman and Rensselaer H. Havens. Townsend Papers.

not without reason, that the whole deal had been planned to enable Eveleth and Bissell to unload some of their stock.[6]

Townsend, now president of the company, conceived the idea of having a thorough examination made of the land at Titusville; if it was found as represented, then he and his New Haven friends would organize a company, assume the lease, drill a well, and monopolize the oil.[7] The New Haven stockholders readily approved the scheme, and he endeavored to interest others; but they would listen to his story, look at him, and shake their heads, remarking: "Oh, Townsend, oil coming out of the ground, pumping oil out of the earth as you pump water? Nonsense! You're crazy."[8] The scoffers did not dissuade him, however, and he began looking for a suitable person to inspect the property.

Boarding at the Tontine Hotel with Townsend was Edwin L. Drake, a man thirty-eight years old and something of a Jack-of-all-trades.[9] He had spent the first nineteen years of his life on farms, first in New York and later in Vermont. With only a common-school education, Drake left home at the age of nineteen for the West. At Buffalo he secured a job as night clerk on a ship plying between that city and Detroit. When the season closed, Drake proceeded to his uncle's farm near Ann Arbor, Michigan, where he worked for about a year. During the next few years he was successively a hotel clerk in Michigan, a clerk in a dry-goods store in New Haven and in New York City. While in the last position he married, and within a short time his wife's failing health caused them to locate permanently in her home town, Springfield, Massa-

[6] Townsend to Brewer, Watson and Company, Jan. 8, 1858. Townsend Papers.

[7] Townsend's Account, 7, 18.

[8] Ibid., 19.

[9] James Dodd Henry, History and Romance of the Petroleum Industry, 127; Bell, History of Petroleum, 5; J. T. Henry, op. cit., 323–325; Carl W. Mitman, "Edwin Laurentine Drake," Dictionary of American Biography (Allen Johnson and Dumas Malone, eds., N.Y., 1928–1936), V, 427–428; Titusville Morning Herald, Sept. 11, 1879, Jan. 28, 1881.

chusetts, where Drake obtained a job as express agent on the Boston & Albany Railroad. In 1849 he resigned to become a conductor on the newly opened New York & New Haven Railroad and moved to New Haven. When his wife died in 1854, Drake broke up the home and, with his only child, went to board and lodge at the Tontine.

While living here, Drake became acquainted with Townsend, talked with him about the Pennsylvania Rock Oil Company of Connecticut, and finally purchased $200 worth of stock.[10] During the summer of 1857 Drake fell ill and, although not prostrated, was compelled to relinquish his position with the railroad. Since he was idle and could obtain a free railway pass, Townsend proposed that he should go to Titusville and examine the land, though the ostensible reason for making the trip would be to perfect the defective title. Thinking that the journey might be beneficial, Drake accepted the offer and in December, 1857, started for Titusville with money that Townsend had furnished.[11] To give the whole affair a pompous turn in the eyes of the frontiersmen at Titusville, Townsend mailed the legal documents and several letters to "Colonel" E. L. Drake in care of Brewer, Watson and Company before Drake ever left New Haven, the title being an invention of Townsend's. Drake has ever since been known as "Colonel" Drake.

On his way west, Drake stopped to see the salt wells at Syracuse, New York, and then proceeded to Erie, Pennsylvania, where he took a stage for Titusville, about forty miles distant. Having been prepared for his coming by the arrival of letters addressed to "Colonel" Drake, the citizens of Titusville provided him with a warm welcome. Drake's legal business required less than three hours, but he had to stay over three days before the next stage ran to Erie. In the interim,

[10] A few months afterwards Drake concluded that Townsend had "pulled" him into the speculative scheme in trying to get himself out, and regretted making the investment. Drake to J. McCarthy, June 28, 1872. Brewer Papers.
[11] Townsend's Account, 7.

he inquired about Oil Creek Township, Oil Creek Lake, Oil Creek, and saw a bottle of rock oil in a local store, visited the principal oil spring, and observed the use of oil for lighting and lubricating purposes in the sawmills of Brewer, Watson

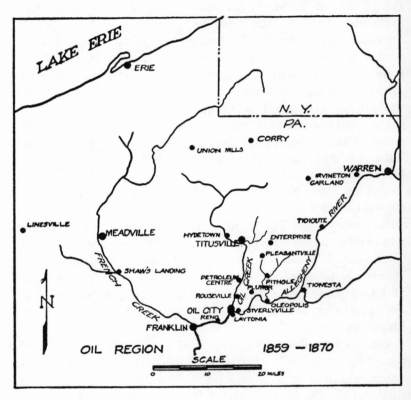

and Company.[12] Fired with an ambition to drill for oil, Drake went on to Pittsburgh to complete his legal business, visited the salt wells at Tarentum, then hurried back to New Haven and enthusiastically told Townsend that he believed

[12] *Titusville Morning Herald*, Sept. 11, 1879; James Dodd Henry, *op. cit.*, 131; J. T. Henry, *op. cit.*, 86; Titusville *Morning Herald*, July 27, 1866; Drake's Account, 3.

oil could be found in large quantities and a fortune made from rock oil.[13]

As a result of Drake's investigation, on December 30, 1857, the three directors in New Haven, constituting a majority of the board, leased the property at Titusville to Drake and E. B. Bowditch, a New Haven cabinet manufacturer, for fifteen years for one dollar and one-eighth of the oil.[14] Townsend and his colleagues believed they had done a splendid thing for the Company in leasing the land to such responsible parties at a figure twenty-five per cent greater than Havens' last and best offer of ten per cent of all oil raised. Moreover, they had confidence in the lessees.[15]

After signing the lease, Townsend received the next day, much to his surprise, an offer from Havens indicating his willingness to lease the land for fifteen years and pay a royalty of twelve cents a gallon.[16] Townsend conferred with other stockholders but, even though it was a better offer, there was nothing they could do, for the lease had been executed and sent to Franklin, Pennsylvania, to be recorded.[17]

At the annual meeting of the stockholders held in New Haven several days later, Eveleth and Bissell first learned of the lease to Drake and Bowditch and "broke out in the most immoderate rage without knowing a word of the Lease" and before any explanation could be made.[18] When the secretary attempted to read the lease, they created such a commotion that he was compelled to stop. Furthermore, Eveleth and Bissell attempted to secure an adjournment. Failing in this,

[13] Townsend's Account, 8; James Dodd Henry, *op. cit.*, 131; Titusville *Morning Herald,* July 27, 1866.

[14] Lease of the Pennsylvania Rock Oil Company to Edwin L. Drake and E. B. Bowditch on Dec. 30, 1857. Townsend Papers.

[15] Townsend to Eveleth and Bissell. There is no date on this letter, but it must have been written shortly after the leasing of the land. Townsend Papers.

[16] Eveleth and Bissell to Townsend, Dec. 30, 1857. Townsend Papers.

[17] Townsend to Eveleth and Bissell, Jan. 14, 1858. Townsend Papers.

[18] Townsend to Brewer, Watson and Company, Jan. 8, 1858. Townsend Papers.

their anger so increased that they withdrew from the meeting. Controlling about two-thirds of the stock, they refused to concur in granting a lease whose terms were less favorable than those offered by Havens. If Drake and Bowditch attempted to begin work, Eveleth and Bissell threatened to secure an injunction. Rather than have the whole transaction brought into court, the directors in New Haven and Drake and Bowditch made a supplementary lease, changing the royalty from one-eighth of the oil to twelve cents a gallon, the same rate as Havens had offered.[19]

With the contract securely tucked away the New Haven capitalists now revealed their hand. They organized the Seneca Oil Company of Connecticut on March 23, 1858, with a capitalization of $300,000, divided into 12,000 shares.[20] Shunning all publicity, they published the Articles of Association as required by law, in an obscure little weekly newspaper of one of the villages of New Haven County. Drake was president and the leading stockholder with 8,926 shares; two other New Haven men, William A. Ives and J. F. Marchal, had 2,680 and 394 respectively. Townsend's name was not listed among the stockholders because, as president of the City Savings Bank, he did not want depositors to know that he had any connection with a scheme which smacked of wild speculation. According to a previous understanding, however, Drake transferred all but 656 shares to other men, all of whom were from New Haven and stockholders in the Pennsylvania Rock Oil Company of Connecticut.[21] Four days after the organization of the company, Drake and Bowditch assigned their lease to the Seneca Oil Company.[22] The directors then elected Drake general agent of the com-

[19] Lease of the Pennsylvania Rock Oil Company of Connecticut to Drake and Bowditch on Feb. 12, 1858. Townsend Papers.

[20] The Articles of Agreement and Association may be found in the Minute Book of the Seneca Oil Company. Townsend Papers.

[21] The stock transfers are recorded in the Minute Book of the Seneca Oil Company.

[22] Transfer of lease, dated March 27, 1858. Townsend Papers.

pany at an annual salary of $1,000, and voted that $1,000 be placed at his disposal to begin drilling for oil at Titusville.[23]

Since he had remarried in 1857, Drake made immediate preparations to move his wife and family to Titusville. They arrived early in May and lived at the old American Hotel until a suitable house could be found. Beyond the ordinary interest created by a stranger's appearance in the village, Drake's arrival and his intent to embark upon the oil business caused no sensation. People along Oil Creek had collected and sold oil for years.

Without boasting about what he expected to do, Drake went quietly to work. As general agent of the Seneca Oil Company in charge of operations, he had been directed to drill for oil, a fact neither understood nor appreciated in Titusville. "In Titusville," declared R. D. Fletcher, "it was all Drake, so far as the drilling of the well was concerned." [24] The fact that none of the stockholders of the company ever appeared in the community gave color to this view. Furthermore, it was not even generally known that Drake held stock in the company. Since the scene of activity was a long distance from New Haven, considerable discretionary authority was, nevertheless, left to him in executing the plan. He started up the machinery abandoned by Angier and hired some men to open up new springs, collect oil, and prospect for coal, salt, and other valuable minerals. A heavy rainfall in May and the first part of June hampered his collec-

[23] Minute Book of the Seneca Oil Company.

[24] James Dodd Henry, op. cit., 156. Jonathan Watson of Titusville seemed to be fully aware of the relationship, however, for he said Drake was "simply an agent for this company." Jonathan Watson, "The City of Titusville," Centennial Edition of the Daily Tribune Republican of Saturday Morning, May 12, 1888, Containing a History of the Founding of the City of Meadville and Settlement of Crawford County and Its Growth and Development During One Hundred Years; an Account of the First Centennial Celebration Held at Meadville, May 11 and 12, 1888, together with Historical and Biographical Sketches of Prominent Men and Events (Meadville, 1888), 165. This quotation is used by permission of the Tribune Publishing Company of Meadville, Pa.

tion of oil; but by the middle of June he was gathering about ten gallons a day, and things began to look "greasy."

Simultaneously, Drake started to dig a well at the site of the principal spring about 150 feet from Upper Mill, which became the headquarters for all persons interested in watching Drake's activities. Shovels, picks, spikes, and a chain could not be purchased in Titusville, and Drake had to get them from Hydetown, Pleasantville, Enterprise, and Erie. In fact, it was frequently necessary to go to Erie and Pittsburgh for machinery and supplies. After several weeks of excavating, the workmen struck a vein of water that drove them out of the well; Drake abandoned the work and decided that it would be cheaper to drill.

Without any practical experience in drilling, Drake went to Tarentum to observe the manner of drilling salt wells and to consult with salt-well owners.[25] While there he learned that many drillers preferred whisky to any other liquor for a steady drink, and frequently a hole was spoiled by a drunken driller, thereby causing a loss to the owner of not only the amount paid for drilling but also the time required to drill another well. Drake concluded that he would not pay for any man's carelessness; and he engaged a driller to drill a round, smooth, straight five-inch hole a thousand feet deep for $1.25 for the first hundred feet, then $1 a foot thereafter, the driller to draw no pay, save enough for board and tobacco, until the well had been completed.[26] If the driller failed to reach that depth through his own negligence or carelessness, he agreed to forfeit all pay.

Upon his return home, Drake ordered a six-horse-power engine and a "Long John" stationary, tubular boiler, the type used by steamers on the Ohio and Allegheny rivers, to furnish power for drilling. He designed an enginehouse

[25] Drake's Account, 5; *Oil City Derrick*, Aug. 27, 1909; Titusville *Morning Herald*, July 27, 1866.
[26] Drake's Account, 5.

and derrick, in which to swing the drilling tools, and ordered the material from Brewer, Watson and Company. On handing the bill to the wood boss, Jonathan Watson, a member of the firm, said that Drake was going to drill a hole through the rock and verify his belief in the existence of a big basin of oil if it took a year. "I have no faith in the project myself," Mr. Watson added, "but I am going to help him through with his venture." [27] By the middle of August, Drake had completed the enginehouse; and the derrick, which had been built lying on the ground near the oil spring, was ready to raise. The derrick was twelve feet square at the base with four timbers thirty feet long, gradually sloping to three feet square at the top. At the appointed time about two dozen men from Upper and Lower Mill and Titusville good-naturedly caught hold and helped raise the structure. All together the work required nearly an hour, and the men viewed it with astonishment when completed. Dubiously shaking their heads and laughing, they called the derrick "Drake's yoke"; the whole enterprise seemed to them "wild and woolly." [28]

In view of past, present, and future expenditures, Drake stood in need of more money. The company had provided him with $500 on April 20 and sent an additional $500 on August 14, but he needed more funds and asked that $1,000 be sent by September 10, 1858.[29] When the stockholders assembled in New Haven for their annual meeting early in September, they immediately voted to send him $500 by the fifteenth of the month and another $500 by the fifteenth of October.[30] For some unknown reason a delay occurred in

[27] James Dodd Henry, *op. cit.*, 158.
[28] Titusville *Morning Herald*, Oct. 23, 1868.
[29] J. T. Henry, *op. cit.*, 89–90; Ledger of the Seneca Oil Company. Townsend Papers. From the time of his arrival in Titusville, R. D. Fletcher, a local merchant, extended Drake liberal credit at his store. Except for one or two items, all of the merchandise seems to have been for Drake's own personal use.
[30] Minute Book of the Seneca Oil Company.

sending the money, but they forwarded $500 on October 30 and the rest on December 30, 1858, making a total of almost $2,000 which the company had sent to Drake during the year.[31] For the same period his expenditures amounted to a little over $800.[32] The stockholders also elected Drake president of the company and a member of the board of directors. Certainly the financial support and his elevation to the presidency constituted a vote of confidence, and there was not the slightest disposition to be out of patience with him for going too fast or too slowly.

Though by the middle of August everything was in readiness for the driller, he failed to appear. After waiting a reasonable time, Drake went to Tarentum to find him, but the man could not be located. Later, it was learned that he regarded Drake as "crazy" and only promised to come to Titusville as an easy way of getting rid of him. Drake contracted with another driller; but unexpected trouble developed at the salt well on which he was working, and he could not come. Since another competent driller could not be found and the season was late, Drake, on the advice of Lewis Peterson, Jr., decided to suspend operations for the winter. Returning to Titusville he installed his boiler and engine and settled down to await the arrival of spring.

Interested in getting an early start, Drake went to Tarentum in February, 1859, to engage another driller, who also failed to appear at the agreed time. Thoroughly discouraged, Drake was on the verge of giving up when a letter arrived from Lewis Peterson, recommending the services of a forty-seven-year-old blacksmith, W. A. Smith of Salina, who intended to quit blacksmithing and go to farming.[33] Smith made the drilling tools and salt pans for Kier and Peterson,

[31] Ledger of the Seneca Oil Company.
[32] Seneca Oil Company account with Colonel Drake from March, 1858, to Feb. 1859. Townsend Papers.
[33] Drake's Account, 8; *Oil City Derrick*, Aug. 27, 1909; *Titusville Morning Herald*, Jan. 8, 1880.

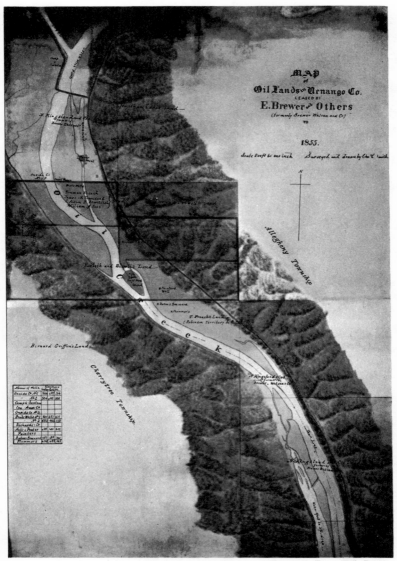

THIS MAP HUNG IN THE OFFICE OF THE
SENECA OIL COMPANY, NEW HAVEN

THE DRAKE WELL IN 1864

Uncle Billy Smith is seated on the wheelbarrow in the foreground.

and if they lost any drilling tools, he always did the "fishing." Deciding to make one last attempt, Drake set out to see Smith. As a consequence of the interview, Smith agreed to come for $2.50 a day and "throw in" the services of his boy. He finished several small jobs of work which took about four weeks, and then he made the drilling tools—the kind that were commonly used in drilling a salt well—for they could be made more cheaply and quickly in his own shop than in Titusville.[34] As soon as Smith was ready, Drake sent a large spring wagon to bring the drilling tools, Smith, his fifteen-year-old son Samuel, and his daughter Margaret Jane. They arrived in Titusville about the middle of May.[35]

Even after Smith arrived there were aggravating delays and loss of time. For some time Drake's men had been at work digging a common well hole and cribbing it with timber; but, owing to the fact that Oil Creek was only 150 feet away and the bed of the stream was higher than the site of the well, water seeped into the hole, forcing the men to work knee-deep in water.[36] Smith rigged up a pump to pump out the water and continued the digging; but with cave-ins and the water running in faster than it could be pumped out, they abandoned the method when down sixteen feet and decided to drive an iron pipe through the quicksands and clay to the rock. From Erie, Drake got some cast-iron pipe in sections ten feet long. With a white-oak battering ram, lifted by an old-fashioned windlass which had already helped in removing the surface dirt, they drove the pipe thirty-two feet to bedrock and, about the middle of August, began to drill with steam power, averaging about three feet a day.[37]

Few people in Titusville paid much attention to Drake's

---

[34] Drake purchased the iron for making the tools, and the cost amounted to $30.80.

[35] Mrs. Smith, her two daughters, and two sons arrived in Titusville in July.

[36] James Dodd Henry, *op. cit.*, 139.

[37] Bell, *op. cit.*, 21.

activities. So many cave-ins and mishaps impeded his work that, as the weeks lengthened into months, the Titusvillians gave less and less thought to the enterprise. Occasionally some one dropped the remark that "Drake was fooling away his time and money." [38] Almost everyone regarded drilling for oil as visionary and sure to prove abortive; there was a complete lack of confidence in the idea, and at times his activities were the subject of ridicule. Even Dr. Brewer, a stockholder in the company which leased the land, seemed to regard the whole affair as a joke. To the young men in Titusville, he gave away cigars, laughingly saying: "Have some on me! They didn't cost me anything. I traded oil stock for them." [39] On another occasion R. D. Fletcher made a business trip to New York City and called on an old friend of Drake's, Babbitt of Babbitt's Schnapps. The latter wanted to know where Drake was and what he was doing. When Fletcher told him, Babbitt laughed long and heartily; Fletcher thought he never would stop. "You don't mean to tell me that Drake thinks he can get oil out of solid rock?" asked Babbitt.[40] Fletcher said that was precisely it, and then Babbitt laughed more than ever.

So much time had elapsed and so many delays had occurred that the enthusiasm of the New Haven stockholders waned; they had already spent $2,000 without any sign of success. After voting to send Drake $500 in April, 1859, they refused to advance any more money. However, Drake had $347 in cash on April 1, and the additional $500 made a total of $847 with which to continue the work.[41] When this was gone Townsend personally continued to bear the expense; but finally even he became thoroughly discouraged and decided to abandon the work. He sent Drake a last remit-

---

[38] Gale, *The Wonder of the Nineteenth Century*, 66–67.
[39] James Dodd Henry, *op. cit.*, 156.
[40] *Ibid.*
[41] Report of the Agent of the Seneca Oil Company for the Quarter Ending Aug. 31, 1859. Townsend Papers.

tance, told him to pay all bills and return to New Haven.[42] To make the situation worse, Drake found himself financially embarrassed during the summer of 1859; so he got his good friends R. D. Fletcher, now a merchant, and Peter Wilson, the druggist, to endorse his note for $500 at a Meadville bank.[43] This loan enabled him to meet his obligations and continue his work.

On Saturday afternoon August 27, as Smith and his sons were about to quit work, the drill dropped into a crevice at a depth of sixty-nine feet from the surface and slipped down six inches.[44] The men pulled out their tools and went home without any thought of having struck oil; they expected to go down several hundred feet more. Late Sunday afternoon "Uncle Billy" Smith, as he became affectionately known, visited the well, peered into the pipe, and saw a dark fluid floating on top of the water within a few feet of the derrick floor. Quickly he plugged one end of a short piece of tin rain spouting, lowered it into the pipe, and drew it up filled with oil. Excited and overwhelmed, "Uncle Billy" sent his boy running to Upper Mill crying, "They've struck oil!" At first the mill hands didn't believe him; but Jonathan Locke, the foreman, stopped the mill, and a dozen or more workmen rushed to see what had actually happened.

[42] Townsend's Account, 10–11. The money arrived either on the day on which Drake struck oil or a few days later.

[43] On account of this incident, the people in the oil region believed that the Seneca Oil Company had neglected Drake, refused to advance expense money and pay the debts. This local feeling was intensified after a representative of the Seneca Oil Company came to Titusville shortly after the completion of Drake's well, and paid for the items charged to the company at Mr. Fletcher's store, but refused to pay any of Colonel Drake's personal charges. The extent to which the Seneca Oil Company and Townsend supported Drake has already been indicated and, even after the completion of his well, the Company sent him $1,118 in September, $400 in November, and $1,118 in December, 1859.

[44] A wide difference of opinion exists as to the day on which Smith struck oil at the Drake well. The best statement is: Smith struck oil on Saturday August 27th; its presence was observed on Sunday the 28th, and the fact became generally known on Monday the 29th.

To their amazement, there was "Uncle Billy" proudly dipping out oil, a larger quantity than they had ever seen.[45] Quickly the news spread, and the dwellers around Upper and Lower Mill and along Oil Creek rushed into Titusville, yelling to everyone they met, "The Yankee has struck oil."

The next morning, when Drake came down to the well, he found "Uncle Billy" and his boys proudly guarding the well, with tubs, wash boilers, and several barrels already filled with oil. Drake got twenty feet of pipe, attached it to a common hand pump, fastened the handle to the walking beam, and began pumping oil. Amid the excitement no one thought of gauging the well. The best evidence indicates, however, that it produced oil at the rate of about eight to ten barrels a day.

Drake seemed pleased to have successfully completed the well, but did not appear greatly excited or wildly enthusiastic; he was not an excitable sort.[46] It is dubious whether he realized the significance of his achievement. Certainly it was beyond the comprehension of the local inhabitants. They could understand the collection of a gallon or two of oil a day; that seemed reasonable, but there couldn't be any possible use for 300 to 400 a day. The people of Titusville, nevertheless, seemed to be very happy over Drake's success, for he had worked long and hard for over a year at his "crazy scheme" and finally had triumphed over a multitude of difficulties.

To judge by the belated recognition of the event and the brevity of accounts, even newspaper editors failed to grasp the significance of what Drake had done. The *New York Tribune* did not mention the affair until September 13, 1859, and then it simply printed a short communication from a Titusville correspondent signed "Medicus." The old *Crawford Journal* at Meadville, Pennsylvania, was the first

---

[45] *Titusville Morning Herald,* Jan. 8, 1880.
[46] James Dodd Henry, *op. cit.,* 154.

newspaper in the oil region to report the strike. Buried away among local items on the inside page, without any caption, the paper carried, on the same day as the *New York Tribune*, a few lines about the event destined to be of supreme importance to the world. The *Venango Spectator* at Franklin on September 21 and the *Warren Mail* on October 1 copied the communication from the *New York Tribune!* Because they believed that the oil would soon be exhausted, the leading stockholders in New Haven prevailed upon the newspapers of that city to refrain from publishing any account of the affair, and almost two months elapsed before they made any reference to it.[47]

In time, the meaning of this momentous event became clearer. Drake had demonstrated in a practical way how petroleum could be secured in greater abundance, and his well served as a textbook for future drillers; he had tapped the vast subterranean deposits of petroleum in the great basin of Oil Creek; and he had ushered in a new industry which provided the world with a cheap, safe, and efficient illuminant.[48] Not only that, but on the eve of a mighty industrial expansion, Drake had opened up a source of unexcelled lubricating oil, an item of utmost importance to the Machine Age.

[47] Townsend's Account, 13.

[48] Quite naturally all the glory fell upon Colonel Drake for completing the first well deliberately drilled for petroleum. At the same time, the New Haven men felt that they had not received any credit for the achievement.

CHAPTER V

Pioneer Wells Along Oil Creek and the Allegheny River

WITH the primitive means of travel and the total lack of communication facilities between Titusville and other places, the rapid spread of the news of Drake's discovery was phenomenal. It seemed as if the entire population of Titusville and the surrounding country had heard the news simultaneously, for in less than twenty-four hours hundreds of people were milling around the Drake well. An eyewitness wrote that the excitement was fully equal to what he had seen in California at the time of the gold rush.[1] Everyone was wild to lease or buy land at any price and drill a well. Because of the location of the Drake well, they believed that the best place to drill was in the lowlands and as near as possible to the water. Consequently, there was a mad rush to secure land near the Drake well and along the Creek.

More alert and quicker than the others, Jonathan Watson jumped on a spirited horse, galloped off down Oil Creek, and leased the farms of Hamilton McClintock and John Rynd. By the middle of September, Brewer, Watson and Company had leased most of the land along Oil Creek where oil springs could be found. Bissell, who had previously arranged to have word flashed to him in New York from the nearest telegraph office, bought all the stock of the Pennsylvania Rock Oil Company that he could buy and then hurried to Titus-

[1] James Dodd Henry, *History and Romance of the Petroleum Industry,* 228.

ville. He boldly leased or purchased farm after farm along Oil Creek and the Allegheny where there were no surface indications of oil. In a month or two, his purchases amounted to over $200,000.[2] Land bordering on Oil Creek was soon taken up, forcing the late-comers to lease or purchase tracts back from it, and within a relatively short time the entire valley, as far back as and even into the hillsides, had been either leased or purchased.

The amount paid for a lease at this time depended upon the size of the tract and locality. For a two-acre lease on Oil Creek near Titusville, a landowner received a $200 cash bonus and one-fourth to one-half of the oil.[3] At less desirable points, owners received a bonus of $25 to $50 and one-third to one-eighth royalty. Many secured leases during the winter of 1859 for one-tenth of the oil and, in some cases, a still smaller percentage, and no bonus; but such advantageous terms could not be made after the spring of 1860, for prices quickly and steadily advanced in the better localities.

While the terms of leases varied, a leaseholder usually agreed to commence drilling at once and proceed with all due diligence to drill a hole, possibly a four-inch one, 200 feet deep unless he found a vein before reaching that depth.[4] In case he failed to obtain oil at the stipulated depth, the leaseholder had the privilege of deciding whether or not he would drill deeper. Some agreed to drill in a second and even a third place if they failed to strike oil. Leases ran for as long a period as the parties might agree; many of them expired in twenty years; some extended to forty, and a few were granted in perpetuity. Uncertainties of all kinds in

[2] J. T. Henry, *The Early and Later History of Petroleum, etc.,* 92; Hayes, "Report of the United States Revenue Commission on Petroleum as a Source of National Revenue," *House Exec. Doc.* No. 51, 39 Cong., 1 sess., 5; *Oil City Register,* Nov. 9 and 16, 1865.

[3] Gale, *The Wonder of the Nineteenth Century,* 22; Leonard to Asahel Pierpont, March 5, 1860. Townsend Papers.

[4] Gale, *op. cit.,* 23.

the terms of these early leases, owing to the haste with which they were drawn, were the seeds for an enormous amount of litigation in succeeding years.

As soon as a tract of land had been leased or bought, the owner selected a spot on which to drill. Guessing tricks and superstitious devices for locating wells everywhere flourished.[5] Many of the operators used the divining rod. With a Y-shaped witch-hazel stick held firmly in both hands in a horizontal position, a seer would walk slowly about the plot. If the loose end happened to be pulled toward the ground by an unseen force, this was the spot on which to sink the well. In a region so full of oil, it naturally followed that many of the wells which diviners located proved successful, and they were in great demand. Other owners relied more heavily upon dreams to locate a site. Still others, with a more prosaic turn of mind, depended upon the sense of smell to lead them to a suitable spot.

Once a place had been selected, drillers followed a procedure similar to Drake's; they erected a derrick and enginehouse, either dug a hole to bedrock or drove an iron pipe, then started drilling. Those who could afford engines and boilers used steam power to drill, but most of the early wells were put down by a slow but simple process called "kicking down a well." An elastic pole, about fifteen feet long, was placed over a fulcrum with the large end fastened to the ground outside the derrick. The other end extended into the derrick, and two or three feet from the end the drilling tools were connected to the pole. Attached to the very end of the pole were stirrups in which two men each placed a foot and pulled down, permitting the drill to drop on the rock. When they loosened their hold, the spring of the pole pulled it back with enough force to raise the drilling tools a few inches. Repeating this procedure rapidly all day long enabled them to drill on an average about three feet a

[5] *Venango Spectator*, March 15, 1865.

day. Though laborious, it provided men of moderate means and strong muscles a cheap method for sinking a well in shallow territory.

In the midst of the wild scramble for land and preparations to drill new wells, Colonel Drake settled down to pumping his well with the complacent conviction that he had tapped the mine.[6] He secured a larger pump, one capable of raising 1,000 gallons a day, and when the well was tubed and a seed-bag put in, it produced twenty to twenty-five barrels a day.[7] With the increased yield, Drake found it exceedingly difficult to procure enough barrels to hold the oil. He collected as many empty whisky barrels as possible in Titusville and Meadville, then built two vats or tanks out of pine planks, each sufficiently large to hold about twenty-five barrels.

On the night of October 7, while Drake was in Erie, "Uncle Billy," thinking that the vat wasn't filling fast enough, went to the tank house to see if the oil had stopped running. Apparently unaware of the inflammability of oil, he carried a lighted lamp, and stepped up the ladder to peer into the vat. Instantly the gas which had accumulated in the top of the tank house ignited and set fire to the oil. Smith jumped to save himself, and within a short time the building, oil vats, 300 barrels of oil, the derrick and house in which the Smiths lived were ablaze.[8]

The next day, while returning from Erie, Drake met an intensely excited man on horseback, who told him about the fire. After listening to the story, Drake very nonchalantly asked, "Did the hole burn?" Furious, the informer galloped off. If angry over Smith's carelessness, Drake did not show it, for the next morning he arrived at the scene of the fire

[6] J. T. Henry, *loc. cit.*

[7] Subsequent references to the production of wells means the amount produced in twenty-four hours, unless otherwise indicated.

[8] *Titusville Morning Herald*, Sept. 11, 1879; Gale, *op. cit.*, 66; James Dodd Henry, *op. cit.*, 141; *Oil City Derrick*, Aug. 27, 1909.

and said pleasantly, "Well, Mr. Smith, I see you've had a frost—a black one, too." [9] With great energy they repaired the boiler, rebuilt the works, resumed pumping early in November, and the well produced about twenty barrels a day. "This may appear incredible," said the editor of the *Venango Spectator,* "yet it is nevertheless true." [10]

Colonel Drake might have leased or purchased any quantity of land and was repeatedly advised to do so, but he rejected all counsel. When several other wells were soon struck, he realized his mistake; but it was too late and, upon the completion of his well, he practically ceased to be a factor in the development of the petroleum industry and others came in to take advantage of his achievement. Like many another enterprising man, "he shook the boughs for others to gather the fruit."

The subsequent history of Colonel Drake is a tragic story. Elected a justice of the peace in Titusville in 1860, he derived an annual income of about $3,000 on account of the rush for legal conveyances and the need for acknowledgments. At the same time, he bought oil for Schieffelin Brothers and Company of New York City, the commissions from this source increasing his income to $5,000. Three years later Drake sold his home in Titusville, left the oil region, taking $15,000 to $20,000, and went to New York, where he became a partner of a Wall Street broker in oil stocks. By 1866 he had lost most of his money through speculation, and his health was greatly impaired. Moving his family to Vermont, Drake sought rest and quietness; but his illness lingered on. Advised to live near the sea, he moved to a friend's cottage near

[9] *Oil City Derrick,* Aug. 27, 1909. This quotation is used by permission of The Derrick Publishing Company, Oil City, Pa.

[10] *Venango Spectator,* Nov. 16, 1869. The Drake well continued to produce twenty to twenty-five barrels a day, but the production gradually declined during 1860. Either late in March or in April, 1860, Colonel Drake completed his second well a few rods from the first; it yielded about twenty-four barrels a day. He drilled a third one near by during the summer, and it produced about twelve barrels.

Long Branch, New Jersey. His funds were now completely gone, and he became the victim of a neuralgic affliction of the spine, which constantly threatened paralysis of the legs. Unable to work, he spent much of his time in an invalid chair and required much attention. His wife not only attended him, kept house, and cared for the children, but took in sewing in order to provide the family with the bare necessities.

While still able to walk, Drake went to New York one day in the fall of 1869 to look for a position for his eldest son. Accidentally he met Z. Martin, a hotel proprietor from Titusville, who was startled to find that Drake walked with great difficulty; that he had only sixty cents in his pocket; and that he wore the same coat that he had worn nine years previously. Martin provided him with a warm dinner, gave him $20, and upon returning to Titusville, reported Drake's condition, of which no one had been aware. Some of Drake's good friends immediately called a public meeting for the purpose of aiding the man who had laid the foundation for so many splendid fortunes. At the meeting, many expressed themselves to the effect that it was a disgrace for the oilmen to allow Drake to suffer; so they appointed a committee to raise a fund to buy a suitable home and provide an income for the colonel. Subscription lists were opened throughout the oil region, and oilmen were solicited; but, although $3,000 was quickly subscribed, the whole project languished partly because of general apathy and partly on account of a controversy over whether Drake deserved anything as he had been the employee of an oil company and had followed its instruction in drilling his well. After considerable difficulty and fifteen months of effort, the committee was able to raise only about $5,000. Disappointed that the oilmen had not responded more generously, Drake's friends succeeded in getting the State Legislature of Pennsylvania to pass a bill in 1873 providing Drake with an annual income of $1,500 during his life; after his death, it should go to his widow as long as she lived.

Meanwhile Drake and his family had moved to South Bethlehem, Pennsylvania. For a few months he improved; then the muscular neuralgia became more severe, and after 1873 he could not walk and could scarcely use his hands. Suffering from the most excruciating pain, he longed for death; but the end did not come for seven years. In November, 1880, Drake died a pensioner of the State.

William Barnsdall and Boone Meade of Titusville and Henry Rouse, a merchant from Enterprise, drilled the second well on the Parker farm, a short distance above the Drake well. Some time in November, they struck oil; but the yield proved to be less than five barrels, and so they resumed drilling. When W. H. Abbott, a merchant from Newton Falls, Ohio, arrived in Titusville in February, 1860, he looked around and finally purchased an interest in this well. A few days later, when it was down 112 feet, oil rushed up and flowed over the top of the pipe at the rate of about ten barrels.[11] Barnsdall's well, as they called it, now became the center of attraction and the "lion of the valley." Its behavior amazed everyone, for whenever they stopped the pump, oil would spout seven or eight feet in the air. One day a party of visitors came to the well and mounted a ladder leading to a platform from which they might obtain a better view. As they stood there marveling at the sight, the well suddenly and violently erupted, throwing oil far above their heads and saturating their clothes. The production soon increased to twenty barrels and dur-

[11] J. T. Henry, *op. cit.*, 95–96; W. H. Abbott, "The Refining of Petroleum," *Centennial Edition of the Daily Tribune-Republican of Saturday Morning, May 12, 1888, Containing a History of the Founding of the City of Meadville and Settlement of Crawford County and Its Growth and Development During One Hundred Years, an Account of the First Centennial Celebration Held at Meadville, May 11 and 12, 1888, together with Historical and Biographical Sketches of Prominent Men and Events* (Meadville, 1888), 164; Gale, *op. cit.*, 58.

ing the first four months, the owners sold 56,000 gallons of oil at a net profit of $13,300.[12]

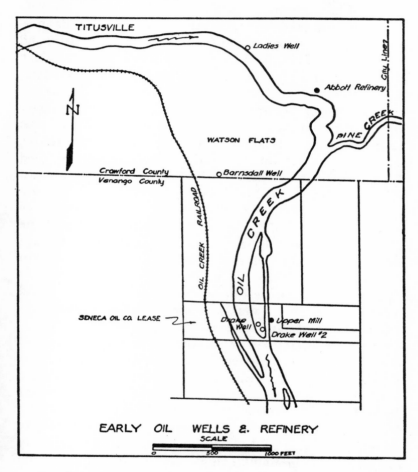

EARLY OIL WELLS & REFINERY

On the opposite side of Oil Creek from the Drake well and half a mile below, David Crossley of Titusville started drilling the third well and struck oil on March 14, 1860;

[12] *Warren Mail*, July 7, 1860; Leonard to Asahel Pierpont, March 5 and 26, 1860, Townsend Papers; *Crawford Journal*, May 15, 1860.

with a pump it produced seventy-five to eighty barrels.[13] On account of the "enormous" yield, the local inhabitants named it the "Elephant" well. The gaping spectators, who had been overawed by the Drake and Barnsdall wells, now flocked to a ferryboat on Oil Creek, paid ten cents fare, and were transported to the opposite bank where they could view, with admiration, the newest sensation. "A splendid thing is the Crossley well!" declared Thomas A. Gale of Riceville. "A diamond of the first water! Enough of itself to silence the cry of humbug; to create a sensation among rival interests; to .inspire hope in many toiling for subterranean treasure, and to make every son of Pennsylvania rejoice in the good Providence that has enriched the state, not only with vast mines of iron and coal, but also with *rivers of oil!*" [14] The editor of the *Crawford Journal* also visited the well and wrote very emphatically: "That there is oil in the bowels of the earth in that locality is beyond question, and that many have struck veins of the rich deposit and are pumping it up in great quantities, is another fixed fact. That others will be equally fortunate none can doubt." [15] If his readers could only see the Crossley well, said the editor, they would exclaim, "The half has not been told us."

The most spectacular well was the Williams well, located in the wilderness about 200 feet from the Barnsdall. On June 23, the driller struck a vein which produced twelve to fourteen barrels.[16] Not content with the yield, the owners decided to drill deeper. After they had gone down three feet, oil burst forth in such a torrent that eighty or ninety barrels were filled in fourteen hours and a large quantity ran out on the ground. The gas, forcing up the oil, made

13 *Crawford Journal*, May 15, 1860; *Venango Spectator*, April 4, 1860; Leonard to Asahel Pierpont, March 26, 1860, Townsend Papers.

14 Gale, *op. cit.*, 59.

15 *Crawford Journal*, May 1, 1860.

16 *Warren Mail*, July 7, 14, and 28, 1860; *Crawford Journal*, July 17 and 31, 1860.

a terrifying noise, and at times oil was thrown ten feet in the air; then it would settle down to a stream two or three feet above the top of the pipe. News of the strike reached Titusville about nine P.M.; a wave of excitement instantly swept over the village, and a tumultuous crowd quickly gathered around the well. One of those present reported, "The coolest nerve could not witness the scene without excitement." [17] Ten barrels an hour were collected throughout the night, and although the flow soon subsided the well continued to be a good producer. People referred to it as the "Fountain" well, and it became "the wonder of the woods."

As the news of Drake's discovery and these other new wells spread, hundreds of people daily poured into Titusville. "It is now the rendezvous of strangers eager for speculation," wrote a close observer in June, 1860. "The capitalists," he continued, "as well as that large class not so rich as ready to venture, are streaming in from all quarters. Here, too, are men who *have toughened their constitutions* in the coal-beds of Ohio, the lead mines of Galena, and the gold placers of California. They barter prices in claims and shares; buy and sell sites, and report the depth, show, or yield of wells, etc., etc. Those who leave today tell others of the well they saw yielding 50 barrels of pure oil a day. . . . The story sends back more tomorrow who must see before they can credit. . . . Never was a hive of bees in time of swarming more astir, or making a greater buzz." [18] Colonel Drake reported in August that Titusville was full of "anxious seekers all determined to make a fortune or bust in the attempt." [19] With a demand for lots and buildings, an appreciation in real-estate prices, a rapidly expanding trade, and a remarkable increase in population, people freely predicted that within

[17] *Crawford Journal*, Sept. 4, 1860.
[18] Gale, *op. cit.*, 13.
[19] Drake to Townsend, Aug. 1, 1860. Townsend Papers.

a very short time Titusville would become one of the principal towns in western Pennsylvania.

The excitement over the discovery of oil in large quantities and the drilling of new wells was by no means confined to the area about Titusville. Simultaneously with the drilling of the Barnsdall's well, Brewer, Watson and Company started putting down a well on the Hamilton McClintock farm at the lower end of Oil Creek, and in November, 1859, they struck oil.[20] By the summer of 1860, at least twelve wells were under way, and a village, McClintockville, had come into existence. On the neighboring farms of J. W. McClintock, A. Buchanan, J. Buchanan, R. Clapp, and others, operators were feverishly at work.

In and around Franklin, capitalists from Pittsburgh and other places purchased or leased scores of sites and prepared to drill. The first to secure oil was E. Evans, a blacksmith in Franklin, who decided soon after the completion of the Drake well to clean out an old salt well at Horsecreek Furnace on the Allegheny, since petroleum had occasionally appeared in the water, especially when low. With the aid of his two sons, Evans made some tools, drilled a short distance, and struck oil, causing a sensation in Franklin. By pumping, they obtained a barrel of oil an hour and equaled the production of any well on Oil Creek at the time. Although Evans's expense of drilling did not exceed $200, he soon received an offer of $100,000 for the well and another to rent it for $5,000 a year.[21] The Hoover and Stewart well, two miles below Franklin, was also finished about the same time; it produced twenty-five to thirty barrels. The successful completion of these two wells in untested territory naturally stimulated further activities, and in August over one hundred wells were being drilled within one mile of the center of Franklin.

[20] *Venango Spectator,* Nov. 16, 1859; *Derrick's Hand-Book,* I, 18.
[21] *Venango Spectator,* Nov. 16 and 30, 1859; *Crawford Journal,* Nov. 8, 1859; Gale, *op. cit.,* 60.

THE PHILLIPS AND WOODFORD WELLS, TARR FARM, 1861

OIL VATS ON THE TARR FARM, 1862

On account of the excitement, this old and dilapidated town suddenly acquired new life. "As might be expected," one writer pointed out in the summer of 1860, "a full tide of visitors is rolling in; claims are taken with avidity. Many are at a loss to decide which will eventually take the lead in the oil trade—Franklin or Titusville." [22] Franklin's population of 936 in 1859 quickly jumped to 1,500 during the summer, and the amount of business increased almost one hundred per cent over the fall of 1859.[23] So great was the town's transformation that a correspondent of a newspaper in Auburn, New York, who spent five days in Franklin during the summer of 1860, hesitated to inform his readers of what had happened lest they charge him with fraud and deception.

Up the Allegheny around Tidioute, the excitement over new wells was at fever heat by August. Oil speculators overran the town, everyone seemed half crazy, and according to the *Warren Mail*, it seemed as if one-half of Warren's population and the rest of creation had gathered at Tidioute.[24] The town had three hotels, and each had three times as many guests as it could comfortably accommodate.

Although several wells had been drilled and many others were in progress, the greatest interest centered in the Hequembourg well.[25] Shortly after its completion in August, there was a groaning and a grumbling in the well and soon a smothered volcano burst forth, throwing oil twenty feet in the air. After it had spouted thirty or forty barrels into the river, the workmen turned its course and saved fifty barrels in an hour. Having filled all available barrels, they plugged the hole until others could be secured and a large vat made. With everything in readiness, they pulled the plug, and the well discharged a barrel a minute; but after flowing forty to fifty barrels it became exhausted, requiring a little time

[22] Gale, *op. cit.*, 14.
[23] *Venango Spectator*, Aug. 1, 1860.
[24] *Crawford Journal*, Aug. 14, 1860; *Warren Mail*, Aug. 18, 1860.
[25] *Crawford Journal*, Aug. 21 and Sept. 18, 1860; *Warren Mail*, Aug. 18, 1860.

for gas to accumulate before commencing again. Never had the spectators seen such a sight!

Over on Tidioute Island, in the Allegheny River, the Ludlow well also started flowing in August.[26] By quick work the owners saved about one hundred barrels and then plugged it. A large vat, big enough for boys to swim in, was quickly built. With scores of people present to see the oil gush forth, they pulled the plug and, to the utter amazement of all, not a drop of oil came out. The crowd roared and yelled, "Humbug!" A. W. Ludlow, one of the owners, perspired, got red, and decided to run a long pole down into the well and stir it. The well sizzled and spluttered a bit, squirted about a gallon of oil, and quit forever.

Despite the failure of the Ludlow well, the island continued to be the center of interest, and valuable wells were drilled during the fall. Owing to the success of different operators, squatters took possession of the shoals and bars in the river and drilled wells. Others believed that the bed of the river might afford an excellent place to drill and within a short time nearly fifty rafts with derricks aboard were at work around the island.[27] Riparian operators, afraid that their wells might be tapped, appealed to the law, and the derricks were pronounced a nuisance and an obstruction to navigation; but drilling continued unabated. As the tension increased, a large raft, manned by about a dozen "tough customers," anchored near the property of one of the most extensive shore operators and a clash seemed inevitable. Both sides made threats, brandished clubs and revolvers, and some pulled off their coats; but they somehow or other avoided a fight. The squatters maintained their position until a flood, more potent than a court order, swept away all but a half-dozen derricks in a single night. Drilling continued

[26] *Warren Mail*, Sept. 15, 1860; *Warren Ledger*, Oct. 31, 1860; *Crawford Journal*, Sept. 18, 1860.
[27] *Warren Ledger*, Oct. 31 and Dec. 5, 1860.

around Tidioute, but the region was soon overshadowed and neglected because of sensational strikes elsewhere.

The production of the pioneer wells for 1859 hardly amounted to 2,000 barrels; but by the end of 1860 a remarkable change had occurred.[28] There were all together about seventy-four producing wells, most of which were on Oil Creek; and they yielded daily about 1,165 barrels.[29] For the entire year, it was estimated that 200,000 barrels had been produced.[30] The drilling of new wells everywhere confirmed the fact that there was an abundance of petroleum; and the amount produced seemed marvelous. At the same time, the production of so much oil had a depressing effect upon the price. In 1859 it sold for seventy-five cents to a dollar a gallon, and little fear was entertained that the price would be materially diminished for years to come. With the greater production, however, and no corresponding increase in the demand, the market value steadily decreased until oil sold for as low as twenty-two cents a gallon in December, 1860.

[28] *Derrick's Hand-Book*, I, 17.
[29] *Venango Spectator*, Nov. 20 and 28, 1860.
[30] *Derrick's Hand-Book*, I, 18.

CHAPTER VI

The Flowing Wells, 1861–1863

THE pioneer wells of 1859 and 1860 produced more oil than anyone had ever seen and created intense excitement throughout the region, but they were nothing compared to the flowing wells drilled along Oil Creek beginning in 1861. The first to startle the inhabitants was located on John Buchanan's farm at the lower end of Oil Creek. The lessees, Little and Merrick, started drilling in the spring of 1861 on the north side of the farm, east of Oil Creek, and back at the junction of the bottom land with a steep hill. About sunset on April 17, 1861, the driller struck not only a heavy vein of oil but a huge gas pocket, and the well commenced gushing at the astounding rate of 3,000 barrels! [1] A workman rushed to Anthony's Hotel, a short distance away, to tell Henry Rouse, one of the lessees of the farm, George Dimmick, the superintendent, and others, who were discussing the fall of Fort Sumter, about the sensational strike. Everyone at the hotel ran toward the well, except Dimmick, who hurried away to secure barrels. News of the strike rapidly circulated, and within a short time at least 150 people had assembled to see this giant gusher.

Having arranged for the barrels, Dimmick ran toward the well and when within twenty rods, a sheet of fire, as sudden

---

[1] *Venango Spectator*, April 24, 1861; J. T. Henry, *The Early and Later History of Petroleum, etc.*, 339.

as lightning, burst forth, followed by a terrific explosion that sounded like the report of a heavy piece of artillery. Instantly, an acre of ground with two wells, oil vats, a barn, and over 100 barrels of oil, were ablaze. The well continued to spout oil high into the air, which fell to the ground, igniting as soon as it fell and adding dense smoke and sheets of flame to the horrors of the scene. Those standing about the well were either stunned or prostrated by the explosion; oil saturated their clothing and most of them became human torches and frantically tried to escape from the fiery furnace.

Rouse was standing within twenty feet of the well when the explosion occurred. Retaining his presence of mind, he started running toward the hill. He had not taken more than a half-dozen steps when he stumbled and fell. Burying his face in the mud to keep from inhaling the flames, he recovered, started up the ravine, fell a second time, exhausted, but two spectators rushed to his aid and dragged him out. His entire body from the head, down the back and legs to the knees, was burned to a crisp. Taken to a near-by shanty and placed on a bed, he suffered the most excruciating pain.

Rouse remained conscious for four hours but showed no sign of his terrible distress. With coolness and precision, he clearly and concisely dictated his will, while some one administered water with a spoon in the middle and at the end of every sentence. After remembering his family and intimate friends, he bequeathed the bulk of his estate to the Commissioners of Warren County, Pennsylvania, the income from which was to be used for roads and to aid the poor. Within less than an hour after completing his will, he died.

Of the eight men talking about Fort Sumter at Anthony's Hotel before the fire, Dimmick was the only one to escape injury, and Rouse the only one to burn to death. All together nineteen persons ultimately lost their lives and the property damage amounted to about $20,000.[2] If the explosion had come an hour later, it would have claimed more victims, for scores of people were constantly arriving at the well.

The well burned for three days and made a magnificent sight. From a pipe six inches in diameter, a solid column of gas and oil extending sixty or seventy feet in the air, burned brilliantly with immense clouds of dense black smoke rising skyward. Hundreds of people came to see the wonder and visit the scene of disaster. The fire was finally smothered with manure and earth and brought under control.

Some believe that the explosion was caused by Rouse's smoking a cigar, but contemporary accounts do not substantiate the charge. On the contrary, it appears that he exercised great care in banning lighted pipes and cigars from the immediate vicinity of the well.[3] The generally accepted view is that the explosion and fire resulted from the gas coming into contact with a boiler at a neighboring well about ten rods distant.[4]

Within a month after the Rouse fire, A. B. Funk, a lumberman from Warren County, Pennsylvania, completed a well on the David McElhenny farm, seven miles below Titusville.[5] Although less spectacular than the Little and Merrick well, it commenced flowing 300 barrels. Skeptics called it the "Oil Creek humbug" and expected that the flow would soon cease; but they waited in vain, for it continued to produce for fifteen months with little variation. Then it suddenly quit and never produced another barrel. By that time, however, it had earned for Funk, who had paid $1,500 for the farm in 1859, about $2,500,000.[6]

The Empire well on the same farm and near the Funk,

[2] *Venango Spectator*, April 24, 1861.
[3] J. T. Henry, *op. cit.*, 343.
[4] *Venango Spectator*, April 24, 1861.
[5] The David McElhenny farm was commonly known as the lower McElhenny farm.
[6] James Dodd Henry, *History and Romance of the Petroleum Industry*, 247.

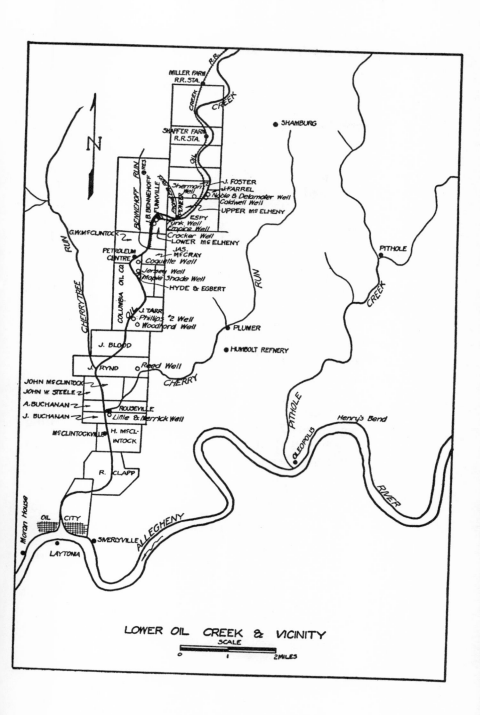

LOWER OIL CREEK & VICINITY

SCALE

0   1   2 MILES

or Fountain, well was more sensational and created fierce excitement. Completed in September, 1861, the Empire started flowing 3,000 barrels! [7] Unable to secure barrels at any price, the owners tried to check the flow, but without success. Therefore, they built a dam around the well and let oil run into the enclosure; but it refused to be dammed and rushed into the Creek. Oil Creek literally became what its name implied, for its surface was covered with oil for miles downstream. The yield simply bewildered the owners. It was too good a thing. With the market already glutted and 3,000 more barrels added to the daily supply, the Empire drove the price of oil down to ten cents a barrel. The owners sold the total product for one month to a Cleveland firm for $500 and not less than 100,000 barrels of oil poured out in the thirty days. [8] After flowing for nearly eight months, the production of the Empire gradually decreased to 1,200 barrels and in May, 1862, like the Funk well, it suddenly stopped.

Within a few weeks after the Empire started flowing, it was eclipsed by a new well on a farm at the lower end of Oil Creek where for years James Tarr had eked out a living. Upon Drake's discovery of oil, Tarr began selling leases. A salt-well driller from the vicinity of Pittsburgh, William Phillips, got a lease, which extended the entire length of the farm along Oil Creek and inland five rods, for one-fourth the oil in barrels. [9] During the summer of 1861 Phillips drilled his first well, and as it commenced flowing 400 barrels he started a second. In October, Phillips Number 2 began flowing 4,000 barrels, the largest well in the region. [10] It finally settled down to 2,500 barrels and continued this yield for months.

Owing to the lack of barrels, the owners of the different

[7] J. T. Henry, *op. cit.*, 227.
[8] *Ibid.*
[9] *Titusville Morning Herald*, Nov. 22, 1871.
[10] *Derrick's Hand-Book*, I, 24.

interests in Phillips Number 2 took their portion of oil by the hour.[11] As boats and barrels were secured or sales made, they would let the oil run two, three, or more hours in each owner's interest. When the barges, barrels, or tanks filled up before the expiration of the allotted time, they let the oil run into the Creek or upon the ground. Many were afraid that the supply of oil would soon be exhausted if something was not done to prevent such waste. Since the lessee was obligated to deliver the landowner's share in barrels and barrels were not available, a stopcock was soon applied and the flow shut off until storage facilities could be provided. They dug holes in the ground for temporary use and then built underground wooden tanks.[12] The latter were eight to twelve feet in diameter, eight to ten feet deep, and walled to the top with pine planks. The top came just even with the ground and to prevent people from falling in, the tanks had to be kept covered. In time, these tanks covered several acres.

About four rods away from Phillips Number 2, N. S. Woodford drilled another well, which began flowing 1,500 barrels in July, 1862. Water from the Woodford soon flooded Phillips Number 2 and materially reduced the flow of oil from the latter. Presently, the peculiar condition developed where neither well would produce oil unless they were both pumped at the same time. The owners then negotiated an agreement whereby both wells should be pumped simultaneously and each would get one-third of the production of the other well.[13]

South of the Tarr farm was the Blood farm, and to the west, the Story, both of which early developed flowing wells. In 1861, the Blood farm had twelve, some of which were very heavy producers. The largest, drilled by the Maple Tree Oil

[11] J. T. Henry, op. cit., 232.
[12] Oil City Derrick, Aug. 27, 1909; George W. Brown, Old Times in Oildom (Oil City, 1911), 92.
[13] J. T. Henry, op. cit., 106, 237; Titusville Morning Herald, Oct. 30, 1865.

Company in October, 1861, flowed 2,500 barrels.[14] The Story
farm had been purchased for $40,000 by some Pittsburghers in
the rush of 1859; they leased plots to various operators and
secured a number of producing wells. Discouraged by the
low price of oil, they decided to change into a joint stock
company and organized the Columbia Oil Company in May,
1861, with a capital of $200,000, divided into 10,000 shares.[15]
Andrew Carnegie was one of the heaviest stockholders, and
the subsequent dividends from oil helped him to erect his new
steel mills. Persuaded by his good friend William Coleman,
Carnegie accompanied him on a trip to the farm about 1862.
At the time, there were on the farm twelve flowing wells,
whose average daily yield amounted to 1,375 barrels.

Coleman proposed to make a lake of oil by excavating a
pool sufficient to hold 100,000 barrels, the waste to be made
good every day by running streams of oil into it, and to hold
it for the not far distant day when, as Coleman and Carnegie
expected, the oil supply would cease. Coleman predicted that
when the supply stopped, oil would bring $10 a barrel, and
then they would have $1,000,000 in the lake reserve. The
suggestion was promptly acted upon, and after losing many
thousands of barrels of oil waiting for the expected day, the
reserve was abandoned owing to the seeming inexhaustibility
of Mother Nature's storehouse.

Probably the best single strike during 1862 was the Sherman
well. Its owner, J. W. Sherman of Cleveland, came to the
oil region in the early days and, like many others, was a man
of limited means.[16] Securing a lease on the Foster farm, he
sank a well with a spring pole. Before reaching the second
sand rock, he exhausted his funds, and as the spring pole
proved powerless, he stopped work. Disposing of a sixteenth

14 Titusville Morning Herald, Aug. 20, 1868; Venango Spectator, Oct. 23, 1861.
15 Crawford Journal, Dec. 15, 1863; Titusville Morning Herald, Aug. 1, 1868.
16 J. T. Henry, op. cit., 101, 224; J. H. A. Bone, Petroleum and Petroleum Wells (Philadelphia, 1865), 26.

interest in the well for an old horse, he resumed work, but two or three weeks of drilling proved too much for "Old Pete." Sherman gave another sixteenth to two men with a small steam engine and once again commenced work. Coal for the engine was not only expensive, but an item for which one had to pay cash; so, in order to buy this, Sherman sold another interest for $80 cash and an old shotgun. He had exhausted his cash and was about to abandon the hole when his drill dropped into a crevice late in March, 1862, and tapped a vein of oil which spouted about 1,500 barrels. For some months the Sherman was the largest on the Creek. It continued to average about 900 barrels daily for nearly two years, and after that was successfully pumped for a long time. During the first year the proprietors made little if any money owing to the low price of oil and the difficulties of getting it to market, but the second year the price improved and the Sherman well brought wealth to its poverty-stricken owners, the total receipts amounting to about $1,700,000.[17]

The striking of such flowing wells in 1861 and 1862 resulted in a remarkable increase in the production of petroleum; it jumped from almost 1,200 barrels a day in 1860 to over 5,000 in 1861. A fabulous supply and a limited demand had the obvious effect of depressing the price of oil. The market condition was also adversely affected by two other factors: first, agents for eastern oil buyers constantly circulated through the region, asserting that they could buy unlimited quantities of oil in West Virginia for a mere song; and secondly, the outbreak of the Civil War caused a panic to sweep over the country, which fell heavily on the oil industry. As a result, the price of crude oil at the wells dropped from $10 in January, 1861, to 50 cents in May, and 10 cents at the end of the year.

With oil almost valueless, owners of flowing wells considered themselves worse off than if they had no oil. Those with

[17] J. T. Henry, op. cit., 225.

wells pumping from five to twenty barrels a day found it most discouraging when an adjoining well spouted hundreds of barrels, flooding the market and making the operation of pumping wells unprofitable. Scores of men with small pumping wells and others with partly drilled wells concluded, therefore, that the oil business had played out, abandoned their leases, and fled in despair, leaving machinery, buildings, wells, in fact, everything.[18]

Faced with an economic crisis and threatened with ruin, some of the landowners and operators of wells drew together and met at Rouseville on November 14, 1861, to organize and take measures to improve the price of oil. One of the producers presented some articles of agreement, the intent of which was to fix a proper valuation on crude oil, regulate the supply and demand, and systematize the oil industry so as to make all business pass through an authorized head and thus furnish accurate information to all persons concerning the market.[19] Considerable discussion followed, and at the next meeting they organized the "Oil Creek Association" to include all operators on Oil Creek and from Tidioute to five miles below Franklin.[20] The plan provided that officers should be elected annually; that an inspector should be elected to regulate the production of flowing wells; that oil should not be sold for less than ten cents a gallon at the wells; and that the proceeds should be paid into a general treasury, where it would be held for the order of the seller, subject to a certain percentage. Further details were settled at subsequent meetings. By January, 1862, the first great combination of producers had been organized, and they refused to sell oil below $4 a barrel.[21]

Since the price of oil remained low during the first half of 1862, business in the oil field stagnated, operators of limited means were either ruined or forced to sell, stopcocks were

18 *Oil City Register*, Oct. 13, 1864; Bone, *op. cit.*, 23.
19 *Venango Spectator*, Nov. 27, 1861.
20 *Crawford Democrat*, Dec. 10, 1861.
21 *Derrick's Hand-Book*, I, 25; *Oil City Derrick*, Aug. 27, 1909.

freely applied, and the daily production dropped from 7,000 barrels to 4,000. To make matters worse, a proposal was introduced in Congress to tax crude and refined oil. Excited and alarmed, the producers held a great mass meeting in Titusville in March, 1862, and petitioned Congress against levying a tax on crude.[22] As finally passed, the Act of July 1, 1862, only imposed a duty of ten cents a gallon on refined petroleum.[23] With the decline in production and the enlargement of the market, due to the increased usage of petroleum at home and abroad, the price advanced to $4 a barrel by the end of 1862, and the condition of the producer improved. With the steady rise in the price of oil, it was not necessary to have a very large flowing well in order to reap a fortune.

Although containing only thirty-six acres, the Farrel farm was probably the most remunerative piece of property of its size in the whole oil region. This barren tract of land was located on the east side of Oil Creek, opposite the Foster farm. The owner, James Farrel, purchased the farm in 1859 for $200, and most people referred to the transaction as "Farrel's sinful waste of money." Early in February, 1863, however, W. A. Caldwell and Company of Pittsburgh completed a well on Farrel's farm, not far from the Sherman, and it flowed 1,200 barrels.[24] A month later, a one-fourth interest in the well sold for $15,000 cash and an eighth interest in the farm for $10,000 cash.

The largest well on the Farrel farm and the best-paying well in the region was drilled by Orange Noble and George B. Delamater, merchants from Townville.[25] They had leased sixteen acres in 1860 for $600 and one-fourth the oil. Using a

[22] *Derrick's Hand-Book*, I, 26.
[23] *The Statutes at Large, Treaties, and Proclamations, of the United States of America* (Boston, 1865), XII, Chap. 119, Sec. 75, 463. The tax did not become effective until Sept. 1, 1862, and then a drawback equal to the amount of the tax was allowed on all refined petroleum exported. The United States collected $237,389.33 on refined petroleum between Sept. 1, 1862, and Jan. 1, 1863.
[24] *Derrick's Hand-Book*, I, 28; Bone, *op. cit.*, 27; J. T. Henry, *op. cit.*, 101.
[25] James Dodd Henry, *op. cit.*, 251.

spring pole, they drilled 130 feet and abandoned the hole. In 1863, however, they decided to deepen it. One day in May, while they were drilling, oil and water suddenly burst forth, rising 100 feet in the air and enveloping the trees and derrick in a dense spray. The gas roared like a hurricane, the ground shook, and oil flowed out at the rate of 3,000 barrels. Boatmen along Oil Creek were notified at once to come and get all they wanted at $2 a barrel. For days the oil ran into the Creek; when the first flow was spent three men, with rubber blankets over their heads and goggles to protect their eyes, worked in the blinding shower attaching a stopcock, by which it was brought under control.

Because of some subterranean connection, the Noble and Delamater well tapped the Caldwell, a few rods distant, and reduced the daily production of the latter from 1,200 to 300 barrels; several wells were completely drained by the gusher. Afraid that they might experience the same difficulty as the owners of the Woodford and Phillips Number 2, Noble and Delamater purchased the Caldwell for $145,000, the highest figure paid for any flowing well up to that time—and plugged it! [26]

With oil at $4 a barrel and the price steadily increasing, Noble and Delamater's daily receipts varied from $12,000 to $45,000.[27] The tremendous flow continued for eighteen months and then declined. All together the well produced over 1,500,000 barrels of oil and netted the owners over $5,000,000.[28] The total expenditures for the lease, drilling, machinery, and labor amounted only to $4,000, so that for every dollar invested more than $15,000 profit accrued in less than two years! [29]

The owners of the Hyde and Egbert farm also had remuner-

---

[26] *Derrick's Hand-Book,* I, 30; *Petroleum Reporter* (Titusville, Pa.) , June 26, 1863.
[27] James Dodd Henry, *op. cit.,* 251; *Derrick's Hand-Book,* I, 32.
[28] James Dodd Henry, *op. cit.,* 254.
[29] *Ibid.*

ative wells. Without any too much cash, Dr. M. C. Egbert of Rouseville and his brother, Dr. A. G. Egbert, purchased the Davidson farm of thirty-nine acres in 1859 and sold leases to different parties.[30] The purchase price and the subsequent expense soon proved embarrassing; so the Egberts sold a half-interest in the land and one-third of the oil to Charles Hyde, a prosperous grocer and farmer, for $2,675. That transaction marked the beginning of a phenomenal prosperity for the owners. In the spring of 1863, the Jersey well began flowing 350 barrels; and it continued with little variation for nine months. During the summer, the Maple Shade well gushed at the rate of 1,000 barrels daily and kept it up for eight or nine months, bringing considerable wealth to the owners.[31] During the life of Dr. A. G. Egbert, its net earnings amounted to $1,500,000. Other large wells were drilled, and every one proved to be an excellent producer.

Although the Caldwell, the Noble and Delamater, the Jersey, and the Maple Shade added to the production, the total amount of oil produced materially decreased during the first part of 1863 because few new wells were drilled, many of the large wells, like the Woodford and Phillips Number 2, were gradually exhausted, and the small wells only pumped from ten to sixty barrels a day. Moreover, Lee's invasion of Pennsylvania in June caused such excitement that there was almost a total suspension of business throughout the oil fields for several days. With the decline in output and the increase in consumption, a marked improvement in the oil business took place. Oil reached $7.25 at the wells in September, buyers swarmed all over the region, and the demand continued to be brisk. The outlook seemed brighter than at any time since 1860.

[30] *Oil City Register*, Nov. 30, 1865; *Titusville Morning Herald*, Feb. 1, 1872.
[31] J. T. Henry, *op. cit.*, 230; *Oil City Register*, Nov. 30, 1865.

CHAPTER VII

Introducing Petroleum to the World

WHILE the wells along Oil Creek and the Allegheny were producing large quantities of petroleum between 1860 and 1863, strenuous efforts were made to market it at home and abroad. Drake had discovered a rich supply of oil, and for a moment it seemed to be of little value. He immediately got in touch with S. M. Kier and W. MacKeown, oil refiners in Pittsburgh, and started shipping oil to them in September, 1859. A month later, Drake contracted with Kier to supply him with such quantities as he desired at sixty cents a gallon delivered in Pittsburgh.[1] Kier promised to sell the oil in preference to any other that should come onto the market, but in case any other appeared in such quantities as to depreciate the value of Drake's, the price paid by Kier was to be reduced. Under this agreement, Drake shipped to Kier from September, 1859, to January 1, 1860, almost $3,800 worth of oil, and for the same period he shipped over $5,000 worth to MacKeown, with whom he had made similar arrangements.[2]

Ferris, the New York petroleum dealer, wanted MacKeown to enlarge his distilling plant and take in the product of Brewer, Watson's well on the McClintock farm, which was then yielding about twenty barrels a day, in order to control

[1] Contract between S. M. Kier and E. L. Drake, dated May 14, 1859. Drake Museum.

[2] Ledger of the Seneca Oil Company. Townsend Papers.

A BARREL FACTORY, OIL, CREEK

Mather Photograph

LOADING OIL AT FUNKVILLE, OIL CREEK

the market for carbon oil and avoid competition.[3] Unless
they did something of this sort, Ferris prophetically predicted,
within twelve months oil would be selling for barely enough
to pay the cost of production. The officers of the Seneca Oil
Company expressed an interest in the plan and offered to sell
to Ferris all oil they had to spare, if a satisfactory price could
be agreed upon, and obligate themselves not to distill or sell
to others for that purpose.[4] In return, they wanted Ferris and
MacKeown to obligate themselves to distill all oil the Seneca
people sold them. Such an arrangement would enable the
Seneca Oil Company to control the crude oil and Ferris and
MacKeown, the refined. The latter's reluctance to accept
either proposition and the completion of new wells made it
difficult to organize the combination; so the scheme to form
the first oil monopoly failed.

In the meantime the officers of the Seneca Oil Company in
New Haven, endeavoring to promote the sale of oil, ordered
100 barrels from Drake. When it arrived the superintendent
of the railroad peremptorily asked that the petroleum be removed at once because of the offensive smell; it was thought
to be unhealthful. Some of the oil was sold to factories in
New Haven, Hartford, and Springfield; and, to introduce the
oil in New York City, the company employed James M.
Townsend's brother to act as salesman.

In order to introduce his product and solicit orders, Drake
made a trip in February, 1860, to Erie, Chicago, Cincinnati,
and Pittsburgh; but he, as well as the New Haven stockholders, found it difficult to sell. Some of the machinists to
whom they sold discovered that it was an excellent lubricant
and recommended it without hesitation to anyone running
machinery. Others, however, had a strong prejudice against
it; they were afraid that the impurities would injure their
machinery and they objected to the odor. But in spite of the

[3] A. Pierpont to Drake, Jan. 30, 1860. Townsend Papers.
[4] *Ibid.*

adverse criticisms Drake did not lose courage. "It takes time and work to introduce it," he wrote to Townsend, "but I shall succeed, I know." [5]

While in Pittsburgh, Drake happened to meet George M. Mowbray, a chemist associated with the wholesale drug firm of Schieffelin Brothers & Company of New York. The meeting was fortunate, for Drake wanted to get his product on the market and Mowbray knew the value of petroleum. Once these two men had been introduced, they started a parley that did not break up until four A.M.; both men loved tobacco, and tradition says that they consumed almost an entire box of choice Havanas during their conversation.[6] A few days later they went to Titusville. Upon showing Mowbray his well, Drake bitterly remarked: "And now, with all the oil obtained—with our entire capital expended—here am I, straitened for the necessary means to introduce it, and the whole of the coal-oil interest dead against me. I have distributed hundreds of barrels, traveled far and near, offering it for one-half the proceeds that may be realized, but have not obtained one cent returns." [7] Mowbray replied, "If I had your oil in New York, I would soon relieve you from this embarrassment." Colonel Drake said that he had 300 barrels of oil at Union Mills, which could be sent to New York at once; so the two men signed an agreement on March 12 for the marketing of oil.[8]

According to the contract, Drake agreed to ship to Schieffelins all the oil, except that sold to those already customers, from the wells of the Seneca Oil Company. In turn, Schieffelins guaranteed all sales, agreed to return all proceeds after deducting the charges; and for this service, they were to

[5] Drake to Townsend, March 11, 1860. Townsend Papers.
[6] *Titusville Morning Herald*, Feb. 20, 1871.
[7] *Ibid.*
[8] Memorandum of an agreement made March 12, 1860, between E. L. Drake and George Mordey Mowbray. Townsend Papers; *Titusville Morning Herald*, Feb. 20, 1871.

receive a commission of 7½ per cent. The charges, however, were not to exceed $1 a barrel. Schieffelins had the privilege of either refining the oil or making arrangements to refine it at a price not to exceed ten cents a gallon; in case they exercised the privilege, they were to receive an additional commission of 2½ per cent. The contract was to be binding for five years, but it could be terminated by either party on thirty days' notice.

Although Drake had arranged outlets through Kier, MacKeown, and Schieffelins, the chief difficulty everywhere in trying to sell petroleum was, as in the past, its disagreeable odor, the impurities in the oil, and its dark, muddy color. Petroleum needed to be deodorized, decolored, and purified before it could be extensively introduced, but there were no refineries, except the small ones of Kier, MacKeown, and Ferris. With a large supply of oil and little demand for it, the Seneca Oil Company decided to go into the refining business; but internal dissension and bankruptcy soon prevented the execution of the plan.[9] Soon after the completion of the Drake well, however, refineries sprang up like mushrooms all along Oil Creek, the Allegheny, at Union Mills, Corry, and Erie.

W. H. Abbott, James Parker, and William Barnsdall built the first refinery in Titusville in the fall of 1860.[10] A large portion of the machinery and appliances, purchased in Pittsburgh, had to be brought by boat up the Allegheny to the point which later became Oil City, thence up Oil Creek. The acids used in this refinery were shipped from Cincinnati, which made them somewhat expensive by the time they

9 Minute Book of the Seneca Oil Company. Townsend Papers.

10 Abbott, "The Refining of Petroleum," *Centennial Edition of the Daily Tribune-Republican of Saturday Morning, May 12, 1888, Containing a History of the Founding of the City of Meadville and Settlement of Crawford County and Its Growth and Development During One Hundred Years; an Account of the First Centennial Celebration Held at Meadville, May 11 and 12, together with Historical and Biographical Sketches of Prominent Men and Events,* 164–165.

reached Titusville. When completed, the refinery consisted of six stills and bleachers, with all the fixtures and tanks under one roof. The first run of oil was made on January 22, 1861, and the yield did not exceed 50 per cent of the crude. Not knowing how to utilize the by-products, they either dumped into Oil Creek or burned all tar and naphtha.

At the upper end of Cherry Run, near Plumer, John E. Bruns and the Ludovici brothers of New York City erected the Humboldt refinery in 1862.[11] The plant covered several acres, employed over 200 men, and, at the time, was the largest in the oil region. It not only was complete in every detail, but was operated more in accordance with scientific principles than any other refinery; so it naturally produced a superior grade of oil. The proprietors shipped a large portion of their product to Europe, where they had extensive connections. Courteous and prompt in business, the members of the firm were deservedly popular. They contributed in a large way to the building up of Plumer, and by their large purchases of oil were of incalculable benefit to the producers.

The most famous of all the early refineries was the Samuel D. Downer plant at Corry.[12] Instead of waiting for the ruin of his newly established kerosene business, Downer started for the oil field shortly after he heard of Drake's discovery in order to find a suitable location for a refinery. At the same time, Merrill and the Atwoods turned their attention to the problem of refining petroleum. After a careful examination, Downer decided to build his refinery at the junction of the Philadelphia & Erie and the Atlantic & Great Western railroads. Here, in October, 1861, he began clearing away the forest, making bricks, putting up buildings, and in less than two years' time he had the largest and finest refinery in the oil region in operation. Built at a cost of $125,000, it had capacity for refining 400 barrels of oil daily. Over 150 men were

[11] *Oil City Register,* Jan. 11, 1866.
[12] *Warren Mail,* June 13, 1863; Titusville *Morning Herald,* Nov. 28, 1865.

regularly employed to operate the works. Great care was taken to prevent explosions; spectators were seldom admitted, and near the furnaces there was an apparatus for throwing the steam suddenly into every room with great force in case of fire. With this well equipped plant and the aid of Merrill and the Atwoods, who were responsible for many notable improvements in the refining process, Downer proved to be an important factor in developing a market for petroleum.

The number of refineries in the oil region rapidly increased. At the end of 1860, fifteen plants had been established, all of them with a small capacity; three years later, sixty-one refineries, with capacities varying from fifteen to three hundred barrels a day, dotted the oil fields.[13] Pittsburgh, which became the earliest important refining center on account of the navigation facilities on Oil Creek and the Allegheny River, had five large refineries in 1860, and sixty in the spring of 1863, representing a capital investment of $1,000,000.[14] They employed 600 men and had a total weekly capacity of 26,000 barrels.[15]

The refining process in the early day was simple.[16] Refiners ran crude oil into retorts, then subjected it to a heat not exceeding 700° F., which caused evaporation. The vapor passed off by means of a worm, in which it was condensed, and from which the product ran into distillate tanks. The first product was naphtha; the next, illuminating oil, and the last, a heavy oil containing paraffin. The illuminating oil was conveyed from the distillate tanks into others and treated with sulphuric acid, caustic soda, and other chemicals, which purified and

---

[13] *Warren Ledger*, Jan. 7, 1863; Kennedy, "Preliminary Report on the Eighth Census, 1860," *House Exec. Doc.* No. 116, 37 Cong., 2 sess., XI, 73.

[14] *Venango Spectator*, Feb. 6, 1861.

[15] *Crawford Journal*, March 17, 1863; *Derrick's Hand-Book*, I, 42; *Oil City Register*, Aug. 25, 1864.

[16] Hayes, "Report of the United States Revenue Commission on Petroleum as a Source of National Revenue," *House Exec. Doc.* No. 51, 39 Cong., 1 sess., 3. The Downer process is described in the Titusville *Morning Herald*, Nov. 28, 1865.

deodorized it. It was then drained off into tanks, ready to be barreled and sold. In this way 75 or 80 per cent of the crude could be converted into illuminating oil.

The cost of refining was not great. A shanty, still, worm, and all, with a capacity to refine five barrels in twenty-four hours, could be erected in 1860 for about $200; a very efficient refinery for $1,500; and for $4,000, one could be made which would purify daily more than the richest well could produce in 1860.[17] The chemicals cost only a trifle, and the charge for operating a refinery ran low—a few hands and a little coal sufficed. All expenses considered, the oil could be refined for five or six cents a gallon.

For illuminating purposes, refined petroleum possessed numerous advantages over camphene or any of the burning fluids. Not only was it cheaper, but the same quantity of petroleum burned more than twice as long as any of the burning fluids; it gave forth a more uniform light; and the wick did not become clogged or crusted over so soon. More important was the fact that rock oil was safer. Many people had found to their sorrow that the name "burning fluid" was no joke. A lamp containing camphene could not be safely carried about in haste or by careless hands; it could hardly be called portable. Though petroleum, when imperfectly purified, proved exceedingly explosive, it could be freed from almost all volatile substances.

Coal oil provided a splendid light; but the blaze from rock oil was larger and higher, and the light was whiter, softer, and brighter. Rock oil emitted a less disagreeable odor than kerosene, especially after the two had been kept for some time; and it was cheaper. Coal oil could be manufactured for twenty-five cents a gallon plus freight, commissions, and other items of expense, while petroleum could be raised from many of the wells for one cent a gallon. Owing to the abundance and natural superiority of petroleum, a coal-oil manu-

[17] Gale, *The Wonder of the Nineteenth Century*, 40.

facturer who visited the oil fields in 1860 predicted, "If this business succeeds, mine is ruined." [18] Many kerosene manufacturers were soon forced into bankruptcy and financially ruined. Others, like Downer, foreseeing the course of events, quickly converted their plants from coal-oil distilleries to petroleum refineries. By the end of 1860, almost all the coaloil distilleries had disappeared and petroleum refineries had taken their place.[19]

The use of oil as fuel did not come until later, but as early as 1864 a United States Naval Board of Engineers began experimenting with petroleum as a substitute for coal on naval steamers.[20] While the results were favorable, the further development of petroleum as a fuel was thwarted by the discovery that when exposed to the air of a confined space at summer temperatures, it gave off, even through the bunghole of a barrel, a gas which, when mixed with atmospheric air, became explosive and detonated with the force of gunpowder. Until a practical remedy for this could be provided, the board reported that it was useless to experiment upon the best form of apparatus for burning it.

It is difficult to estimate the amount of refined petroleum annually consumed in the United States after 1859. Some idea may be gained, however, from the revenue derived by the United States from the tax on refined petroleum. For the last four months of 1862, the government received $237,389.33; in 1863, $1,179,276.21; in 1864, $2,255,328.80, and in 1865, $3,047,213.[21]

Simultaneously with the expansion of the home market, petroleum was introduced into Europe. Carrying samples of crude and refined oil, Charles Lockhart of Pittsburgh went

[18] *Ibid.*, 54.

[19] James Dodd Henry, *History and Romance of the Petroleum Industry*, 81. The refined petroleum used for illuminating oil soon came to be commonly called kerosene or coal oil.

[20] Secretary of the Navy, *Report for 1864*, 1096.

[21] Hayes, *op. cit.*, 33. From Sept. 1, 1862, until June, 1864, the rate was ten cents a gallon on refined petroleum; after June, 1864, it was twenty cents.

abroad in May, 1860, and called the attention of some of the European merchants to the value of petroleum. During the same year the officers of the Seneca Oil Company sent a sample of oil to Havre, France, with C. H. Townsend, a brother of James M. Townsend and captain of a steamer plying between New York and Havre, to have it analyzed by A. Gelée, a French chemist. After the analysis had been completed, Gelée said to Townsend, "If that oil can be gathered in quantity enough, its illuminating and lubricating qualities are such that for those purposes, it will revolutionize the world." [22] In January, 1861, William Reynolds and J. J. Shryock of Meadville, Pennsylvania, sent twelve barrels of oil to London to James McHenry, the chief contractor for the Atlantic & Great Western Railroad. He distributed samples among the members of the Board of Trade in London and sent some to chemical experts in Paris. "The examination of the oil," McHenry wrote from Paris in June, 1861, "shows a splendid result for the purpose of fabrication of gas. . . . The Gas Company considers it better than anything they know of for the fabrication, and all depends on the cost of oil placed in New York or London. . . . If the oil comes cheap the trade will be fabulous. . . ." [23]

A. W. Crawford, United States Consul at Antwerp, had been at his new post only a month when he reported on October 30, 1861, that "some of the numerous oil springs of the Allegheny and Kanawha Valleys might profitably let their lights shine in this direction." [24] Having been familiar with the properties of Pennsylvania oil, Crawford called the attention of Antwerp merchants to its superiority over colza oil both in economy and in beauty of light; but their strong prejudices required drastic action. At his own expense and

[22] Townsend's Account, 12; Gelée's written analysis is dated Jan. 10, 1861.
[23] John Earle Reynolds, *In French Creek Valley*, 75. This quotation is used by permission of the author.
[24] Secretary of State, *Report on the Commercial Relations of the United States with Foreign Nations, for the Year Ending September 30, 1861*, 176.

risk, Crawford had a small amount of refined petroleum forwarded from New York.[25] Distributing it among the merchants, he soon got favorable results; and petroleum became the leading article of commerce.

When A. J. Stevens, newly appointed United States Consul to Leghorn, Italy, arrived in that city in December, 1861, he found that no effort had been made to introduce petroleum there; in fact, nothing was known of this new "Yankee invention." [26] Fortunately, Stevens had carried with him a sample of petroleum and some lamps, the first to be seen in this line of trade, and his results were most encouraging.

At the request of some New York merchants, W. W. Murphy, the United States Consul at Frankfort, first introduced petroleum into Germany in 1861.[27] Several obstacles, however, prevented it from coming into general use: gas companies opposed it as a formidable rival; insurance companies opposed it on account of its great inflammability; cautious local authorities imposed drastic regulations upon its use; and German railroad companies prohibited its transportation in their cars on account of the odor and danger.[28]

The first petroleum from the United States to be delivered in England evoked an astounding amount of hostility from the press, members of public bodies, and especially coal-oil manufacturers, who saw American petroleum sold in British territory cheaper than coal oil could be manufactured.[29] Afraid that it would destroy their business, they influenced Parliament to levy a tax of one penny a gallon on all petroleum imported from the United States. The opposition also widely proclaimed the highly dangerous qualities of petroleum, and

---

[25] James Dodd Henry, *op. cit.,* 216.

[26] Secretary of State, *Report on the Commercial Relations of the United States with Foreign Nations for the Year Ended September 30, 1865,* 453.

[27] Secretary of State, *Report on the Commercial Relations of the United States with Foreign Nations for the Year Ending September 30, 1861,* 433.

[28] *Ibid.*

[29] James Dodd Henry, *op. cit.,* 268, 310,

in view of the Rouse fire it is not surprising that a genuine fear everywhere accompanied the introduction of petroleum. Furthermore, the press ridiculed the new industry. One writer said, "The whole Atlantic and Great Western Railway smells like a leaky paraffin lamp, and unless some means can be discovered of overcoming the miasma an American and Canadian will be detected in society by his scent as easily as musk deer or a civet cat." [30]

The foreign demand began to assume the proportions of a heavy and regular business in 1862. Such tremendous progress was made in expanding the market that the London *Times* prophesied that the value of this trade might even approach that of American cotton.[31] Petroleum was rapidly winning favor in France, England, and other countries and was destined to supersede the use of candles, rapeseed oil, and other illuminants. Owing to Crawford's efforts, 1,500,000 gallons were sold in Belgium in 1862, an amount sufficiently great to threaten the complete destruction of the growth of rape and linseed.[32] At Leghorn, petroleum was beginning to be used among all classes of people, for they found it was considerably cheaper and gave a better light than olive oil.[33] From its central position and railway connections with the interior, Leghorn became a distributing point for a large section of the country, especially to the north and east of Florence.

During 1862, petroleum from the United States was introduced into Russia. Though large quantities had been discovered near the Caspian Sea and in other parts of Russia, there was only one refinery, and its product sold for a dollar a gallon. A great change in the fluid used for lighting took place in Russia in 1863—a matter of great importance in a

[30] Quoted in James Dodd Henry, *op. cit.*, 268.
[31] Quoted in *Venango Spectator*, Jan. 8, 1862.
[32] Secretary of State, *Report on the Commercial Relations of the United States with Foreign Countries for the Year Ended September 30, 1863*, 272.
[33] *Ibid.*, 487.

latitude where, in winter, there were scarcely six hours of daylight. Formerly tallow had been wholly relied upon; now kerosene was fast becoming universal, and American lamps proved to be the most popular.[34] "The people are becoming accustomed to it," the United States Consul at St. Petersburg significantly reported in December, 1863, "and they will not do without it in the future. It is, therefore, safe to calculate upon a large annual increase of the demand from the United States for several years to come." [35]

Obviously, the advantage of petroleum over all other burning fluids was becoming thoroughly appreciated. By the end of 1863, Antwerp enjoyed the distinction of being the largest petroleum emporium in Europe.[36] Petroleum constituted the principal article of import from the United States, and no port surpassed its facilities for warehousing and keeping oil. Bordeaux in 1864 adopted kerosene in preference to gas for lighting the new public gardens, and the United States government observers predicted that its consumption would increase as the French became better acquainted with its qualities. Moreover, a marked increase in the consumption of petroleum was noted in southern Germany. Of its value to Germans, the United States Consul at Hamburg wrote: "In a country where tallow, wax, common oil, and other substances, used for dispelling the darkness of the long winter evenings and the gloom of the short winter days of this climate, are so very dear, it is hard to imagine how the common people got along before petroleum was discovered." [37]

Remarkable, indeed, had been the introduction and use of petroleum from the United States in foreign lands. Between 1860 and the end of the Civil War, the amount of petroleum

[34] Secretary of State, *Report on the Commercial Relations of the United States with Foreign Nations for the Year Ended September 30, 1864*, 423.
[35] *Ibid.*
[36] Secretary of State, *Report on the Commercial Relations of the United States with Foreign Nations for the Year Ended September 30, 1865*, 238.
[37] *Ibid.*, 430.

exported by the United States rose from nothing to $15,727,881; and it ranked sixth in our exports, exceeded only by gold, corn, tobacco leaf, wheat, and wheat flour! [38] Petroleum now constituted at least one-twentieth of our export trade. Great Britain was our best customer; France, second; Belgium, third; Hamburg, fourth; Holland, fifth; and there was scarcely a country in the world to which we did not ship petroleum.[39]

[38] *Ibid.*, 110–113.
[39] *Ibid.*, 32–33.

CHAPTER VIII

Transporting Petroleum to Market

THE facilities for shipping oil from Oil Creek prior to 1862 were crude and inadequate, except when the water in Oil Creek was high enough to permit flatboating. Highwater stages averaged less than six months a year, and during the rest of the time oil had to be hauled to different shipping points. The nearest railroad stations to Titusville and the oil region in 1860 were Corry, Union Mills, and Garland, about twenty to twenty-five miles to the north. The Philadelphia & Erie Railroad (now incorporated into the Pennsylvania Railroad) served all of these towns and connected them with the seaboard. The Atlantic & Great Western Railroad (now a part of the Erie Railroad), whose tracks crossed those of the Philadelphia & Erie at Corry, also provided direct service to the eastern markets.

Prior to 1862, about 6,000 teams were regularly engaged in hauling oil to these shipping points.[1] No such transportation service had ever been seen in the United States, except in the Army. Referring to teaming through Titusville, one newspaper editor said, in 1860, "The wagons are all loaded with oil, going, going, going, all the time to the railroad, and yet this thing is only in its infancy."[2] Later, more than 2,000 teams, by actual count, passed over Franklin Street bridge in

[1] John J. McLaurin, *Sketches in Crude-Oil* (Harrisburg, 1896), 262.
[2] *Crawford Journal*, May 1, 1860.

Titusville in one day.[3] It was not uncommon to see a solid line of teams a mile or more in length on the roads leading to Union Mills, Corry, and Garland.[4] From dawn until dark, a procession of teams and wagons slowly moved along without any interruption, seven days a week. From 300 to 500 barrels a day were hauled in the fall of 1861 to Shaw's Landing, about thirty miles away on the Meadville and Franklin Canal.[5] Furthermore, 600 to 1,000 barrels daily passed through Meadville for Linesville on the Pittsburgh & Erie Railroad. The *Crawford Journal* at Meadville reported in January, 1862, "Immense trains of wagons are now upon the road carrying oil from the various regions of Greasedom through this place to the Railroad at Linesville."

The horses suffered outrageously from severe lashings, brutal treatment, and excessive work. There were no roads running to the wells, and so teamsters simply pulled down fences and drove through fields wherever possible. In wet weather, the so-called roads were usually full of bottomless mudholes. The oil spilled from passing wagons mixed with the mud to keep the mass a perpetual paste, which destroyed the capillary glands and hair of the horses. On account of this, many of them had no hair on their bodies below the neck and long wagon trains of these hairless animals always excited the curious visitor.[6] Many of the wagons dropped into the mudholes above the axles; horses sank to their bellies or higher; and many of them, falling into the batter, were left to smother. Hundreds of dead horses could be seen along the banks of Oil Creek; some died a natural death, others died from overexertion, trying to pull a heavily loaded

[3] Titusville *Morning Herald*, July 20, 1865. One observer says that at one time there were no fewer than 4,000 teamsters in Titusville alone (Titusville *Morning Herald*, May 25, 1869).
[4] Alfred Wilson Smiley, *A Few Scraps, Oily and Otherwise* (Oil City, 1907), 54.
[5] John Earle Reynolds, *In French Creek Valley*, 300.
[6] McLaurin, *op. cit.*, 263.

wagon out of a mudhole or towing an oil barge up the Creek.[7] The unfortunate teamster would secure another horse; but, having lost his place in line, he was delayed two or three hours, for he had to wait and take up his position at the end of the line. In case his wagon broke down, the driver merely dumped out the load into the mud or left the barrels on the bank and went on, leaving the oil to be taken by anybody who thought it worth stealing.

The customary load was five or six barrels to the wagon, and the charges for hauling naturally depended upon the distance, the condition of roads, and the time of year. However, early in 1862, teamsters received $2.50 to $3 a barrel for delivering oil to Meadville, $3.50 to $4 to Linesville, and about $2.50 to Corry, Union Mills, and Garland. In many instances those on the Tarr and Blood farms paid as high as $5 a barrel in 1861 and 1862.[8] The unprecedented demand for teams led teamsters to ask exorbitant prices—and they got them. It exasperated the producer, however, for two reasons: the price of oil was low because of the flowing wells; and it cost more to transport a barrel of oil from Oil Creek to Union Mills than from Union Mills to New York City.[9]

The only alternative to this expensive and slow method of transporting oil was the pond freshet, which lumbermen had used for years to raft their logs down Oil Creek when the water was too low to permit navigation. To create the pond freshet there were at least seventeen sawmills with dams on the principal branches of Oil Creek, some of which were as much as ten miles above Titusville. Through a system of floodgates, the water could be held until a sufficient quantity had backed up; then it was let loose, thereby making a stage of water below sufficient to float logs down the Creek. After the pond freshet passed, the cuts in the dams

[7] "Oil Transportation," *Petroleum Age*, VII (1888), 34–36; Titusville *Morning Herald*, July 25, 1865.

[8] *Venango Spectator*, Jan. 8 and 15, 1862.

[9] The freight on a barrel of oil to New York in Jan., 1862, was $2.40.

were closed, the water collected, and the mills resumed sawing and grinding until the next one.

Because it was cheaper and quicker, though fully as hazardous as or more so than teaming, the oilmen appropriated the idea of the pond freshet and appointed a superintendent to make all the necessary arrangements. Some of the mill owners were glad to sell the water and the use of their dams because they made more money from the pond freshet than from sawing logs; others were reluctant to have their water and dam used because it meant shutting down the mill from twelve to forty-eight hours; and often others would not cut their dam for any price, especially in dry weather. The mill owner's charge for the use of his dams, the salary of the superintendent, and the services of two men to cut the dams made the cost of each pond freshet $100 to $400.[10] At no time in 1862 did the expense exceed $110, but as the demand upon the mills for lumber increased, mill owners grew more reluctant to part with the water and advanced their price. For the use of the Kingsland dam, the largest one, the owner charged $55 in 1862, but $200 in 1863. Others raised their prices, but not in the same proportion.

To defray the expenses, agents traveled down Oil Creek immediately prior to a pond freshet to ascertain the number of boats ready, collected a toll on each barrel from every boatman, and paid the mill owners in advance. The toll depended upon the amount demanded by the mill owners for using their dams and the quantity of oil to be shipped: in 1862 it amounted to two cents a barrel; and in 1863, owing to the increased expense, a dry season, and a smaller amount of oil to be run out, it was four cents.[11] Shipments of no fewer than 10,000 barrels were required, therefore, to meet the expenses; and, as the cost increased, shipments of 20,000 to 30,000 barrels were necessary.

[10] *Derrick's Hand-Book*, I, 34.
[11] Pennsylvania passed a law in March, 1863, limiting the charge to three cents a barrel.

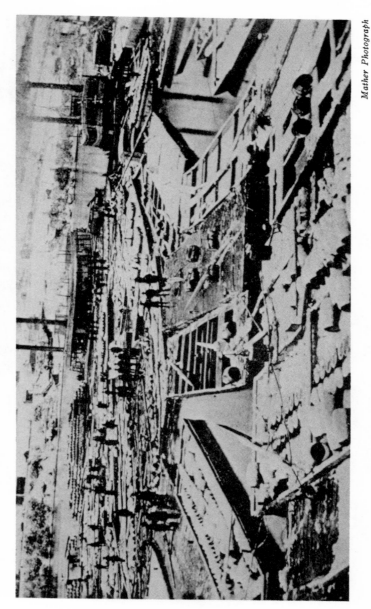

OIL BOATS AT THE MOUTH OF OIL CREEK, 1864

Mather Photograph

PACKET BOAT, OIL CREEK. 1863

During the busy season, pond freshets were usually provided twice a week. The superintendent set the date and notified shippers and boatmen well in advance, so that they could overhaul their flatboats and tow them to a point on the Creek to be loaded. The boats were of all sizes and kinds; most of them would hold 700 or 800 barrels, and they carried the oil either in bulk or in barrels. When oil prices were low and barrels scarce, they usually shipped it in bulk. After all boats had their cargoes aboard and the toll had been paid, word was sent up the Creek "to pull the shoats."

The superintendent and his men would then go to the dams above Titusville and, about midnight, commence cutting.[12] Waiting until the water from the first commenced running over the next dam, they would then cut the second. This process was repeated until they got to the Kingsland dam, three miles below Titusville and a mile below the Drake well. When it was cut, they had turned loose in one body at one time, in some cases, the water from seventeen dams, which caused a rise in Oil Creek of twenty-two to thirty inches above the highest rock on the swiftest riffle. It was important for the dams to be cut according to a schedule: any irregularity would prevent the water from reaching the last dam in a body and therefore from rising to the proper depth, so that the first boats would run aground upon the bars.[13] Those following would have no means to check their headway and no escape from crashing into those on the bars.

A cool breeze was the first sign of the freshet's approach and the swirling waters soon followed. The excitement now began, for life and property were endangered. Expectant boatmen stood ready to cast off their lines when the current was precisely right, a matter which required good judgment. Inexperienced boatmen generally cut loose their boats

12 *Warren Mail,* Jan. 24, 1863.
13 *Titusville Gazette and Oil Creek Reporter,* June 26, 1862.

upon the first rush of water, only to be grounded and battered into kindling by those coming later. An experienced boatman waited until the water commenced to recede, then cut loose his lines, throwing himself upon the mercy of a swift current.

A pond freshet afforded a most unusual sight, for here were 150 to 200 flatboats, little and big, loaded with 10,000, 20,000, or 30,000 barrels of oil, either barreled or in bulk, floating along endways and sideways on a rushing flood and wildly fighting their way down Oil Creek, which was only twelve rods wide and very crooked as it wormed its way through steep hills. It required all the skill and strength of some 500 boatmen to avoid collisions with other boats, the rocks, and other obstructions. At several points along Oil Creek, formidable obstacles tested the skill of even an experienced boatman; one was the pier at McClintock's bridge; another was a pier a short distance below to support the machinery of a well in the middle of Oil Creek; and the third was Forge dam, through which there was only a narrow passage for boats.[14] If a boat got crosswise of the Creek at any one of these points, a jam occurred, and bulk boats, built of very light timber, would be easily crushed and the oil spilled into the Creek. If the oil was in barrels, the boat sank, the barrels floated off, and the owner rarely recovered all of them.

If the boats successfully passed these obstacles, they soon approached Oil City, where Oil Creek emptied into the Allegheny River. The cry of "Pond freshet!" brought out the entire population of the town; it was a gala occasion. On the bridge across the Creek and along the shore hundreds of people were on hand to witness what was locally considered to be one of the wonders of the world. Even now, disaster might overtake the fleet. Often a boat struck a rock a few rods above the Oil City bridge, swung broadside against

14 *Warren Mail*, Jan. 24, 1863.

the middle pier of the bridge, and either broke in two or blocked the channel.[15] Every man on the bridge and shore knew exactly what should be done, and as other boats approached there was no lack of directions and orders from ashore and no scarcity of expletives and brimstones from the men on the boats. Boatmen could see the danger and made frantic efforts to save their property; but on they came, crowding, bumping, and crashing against the boat at the pier, making a report like the discharge of artillery. Splinters flew in all directions and the work of weeks disappeared in a few moments. Then came the vanishing of the waters as they hurried onward and only a high pile of twisted timbers and a mass of greasy wooden splinters was left. Shippers became accustomed to such disasters, however, and took the matter more coolly than one suspects, for there was no place where fortunes were won or lost with more nonchalance than in the oil region.

Oil from overturned or demolished boats floated into the eddies below the mouth of Oil Creek and belonged to whoever dipped it up. Since so much oil went to waste, people improvised small dams near the old Moran House and gathered it.[16] As much as $900 was made in this way in an afternoon by one person when the price of oil was $10 a barrel. Oil was so plentiful on the Allegheny River after one of the disasters that many persons leased land between Oil City and Franklin for the purpose of throwing out booms and taking up oil as it came downstream.[17] An incredibly large amount was collected at these points.

If the fleet passed the bridge successfully and did not lodge on the gravel bar at the mouth of Oil Creek, the boatmen soon had it anchored in Moran's eddy at Oil City. Generally speaking, it was very common for 150 to 200 boats to

[15] *Venango Spectator*, May 21, 1862.
[16] *Ibid.*
[17] *Titusville Gazette and Oil Creek Reporter*, June 26, 1862.

run out 10,000 to 20,000 barrels of oil without serious mishap. But no matter how successful a pond freshet might be, it always involved a heavy loss of oil; a third of it was lost by leakage before the boats started, and another third was lost before it reached Pittsburgh.[18]

After a successful pond freshet, Oil City presented a lively scene. Shippers were busy paying off boatmen; laborers from up along the Creek purchased a new stock of provisions; and all was bustle and business.[19] Men dripping with the oleaginous fluid stood talking on the sidewalks; hotel lobbies were filled with greasy men; the whole atmosphere reeked with the smell of oil; and everywhere, oil was the principal topic of conversation.

While the fleet lay at anchor, a sudden freshet, the breaking loose of an ice gorge, or fire occasionally endangered it. An ice gorge broke loose and came down Oil Creek in December, 1862, crashing against 350 boats with 60,000 barrels of oil aboard, and destroying over half the boats and oil.[20] The loss amounted to over $350,000. At another time, a fire broke out at two A.M. on a flatboat lying at anchor among 300 or 400 others in the eddy below Oil City.[21] The fire originated from a lantern left burning on the edge of an open bulk boat. A watchman, hurrying to reach it, stumbled over an oar, knocked the lantern into the oil, and immediately an explosion occurred, which threw burning fragments into other boats. When a portion of the fleet moored in the eddy was cut loose, the boats swung across the river and communicated the flames to a number of barges on the opposite side. From the lower end of the eddy to the east bank, the river was bridged with oil boats, which, in an incredibly short time, were a mass of smoke and flame; over fifty-two

---

[18] Andrew Carnegie, *Autobiography* (John C. Van Dyke, ed., Boston, 1924), 138.

[19] *Warren Mail*, Jan. 24, 1863.

[20] *Derrick's Hand-Book*, I, 27.

[21] *Venango Spectator*, May 13, 1863; *Warren Mail*, May 30, 1863.

boats were adrift and on fire. At six A.M. a large boat in flames was sighted coming down the river toward Franklin. Despite every effort to land it, the boat continued downstream. Passing under the bridge over the Allegheny, the flames shot to the top of the roof and instantly enveloped the bridge in fire. Within thirty minutes the entire structure had been destroyed. Altogether, between 75 and 100 boats burned, containing from 8,000 to 10,000 barrels of oil. According to the editor of the *Venango Spectator*, it was the worst calamity that had ever visited the oil region.

Usually the morning after a pond freshet the boatmen started to Pittsburgh with the oil in the same boats, or else they transferred it to larger and stronger barges, which quickly came into use on the river after the discovery of large quantities of oil. As early as April, 1860, the steamer *Venango* carried the first load of petroleum to Pittsburgh, and within two years there were fifteen steamboats and towboats plying between Oil City and Pittsburgh, each having an average capacity of about 800 barrels.[22] They averaged about three trips a week when the river was in good navigating condition and the towboats two.

In November, 1861, Jacob Jay Vandergrift of Pittsburgh started the bulk boat business when he towed two large coal boats with 4,000 empty barrels to Oil City with his steamer, the *Red Fox*.[23] While delivering the barrels, Vandergrift bought 5,000 barrels of oil. As he was trying to figure out a way to get his oil to Pittsburgh, the first boat to carry oil in bulk arrived from Oil City. After inspecting it at Allegheny City, Vandergrift believed that this would be a cheap and feasible means for transporting oil; so he contracted with a boat builder to build him twelve boats, eighty feet long, fourteen feet wide, and three feet deep, each with

[22] *Crawford Journal*, April 29, 1862; *Warren Mail*, April 19, 1862; *Venango Spectator*, April 11, 1860.

[23] James Dodd Henry, *History and Romance of the Petroleum Industry*, 259–260.

a capacity of 400 barrels. Upon their completion in the spring of 1862, Vandergrift inaugurated a highly profitable barge business. The freight charges from Oil City to Pittsburgh ran from twenty-five cents to three dollars a barrel; they depended upon the weather, the stage of water, the demand for oil, and whether oil was transported by steamer or bulk boat.

To get freight up Oil Creek and oil down, flatboats pulled by four horses hitched abreast were commonly used.[24] With the bed of the stream even and covered with loose flat shale rock, the horses could easily pull the boat in midstream, although the water might be up to their bellies. In winter, it was anything but pleasant for the horses, their bodies partially clad in icy coats of mail and their tails looking like bunches of icicles. Few long survived the frightful experience of three or four hours a day in the icy water without any rest. Drivers scarcely felt the loss, for a single trip would realize more than enough to buy another horse or team. The flatboats held from 80 to 100 barrels of oil on which the charges ranged from seventy-five cents to a dollar a barrel; and the rate on other freight was in proportion.

Several scows under canvas, drawn by one or two horses, provided passenger service between Oil City and different points on the Creek.[25] A short blast from a tin horn, a reedy twang like the drone of an angry bee, gave notice of the packet's approach. The captain steered the ship, blew the horn, and kept watch for passengers on shore. Spying a passenger, the captain would yell to the man leading the horses, "Ground her," or "Fetch her to." All along Oil Creek the packet stopped like a milkwagon. If the water proved sufficiently high, the horses might be unhitched from the boat upstream, and the packet allowed to float down with the current to Oil City. Otherwise, they pulled it. Nearing Oil

---

[24] Bone, *Petroleum and Petroleum Wells*, 70.
[25] *Warren Mail*, Dec. 9, 1865; Titusville *Morning Herald*, April 14, 1866.

City, the crew's excitement frequently grew tense, for they would overtake the Rouseville stage headed for Oil City. Using the whip like a flail and rushing on without any fear of a wreck, the captain would dip his horn in the water, as Oil City came into sight, in order to give it a clearer note and crow a shrill tin crow like a triumphant gamecock.

No railroad connection was made with any of the towns in the oil field until 1862. Within a short time after Drake's discovery, however, a group of capitalists headed by Thomas Struthers of Warren, Pennsylvania, organized and incorporated the Oil Creek Railroad Company. The charter authorized the company to construct a railroad from any point on the Philadelphia & Erie Railroad to Titusville, thence along Oil Creek to Oil City and Franklin.[26] In order to take advantage of the facilities offered by both the Philadelphia & Erie and the Atlantic & Great Western, the directors built a broad-gauge track in 1862 from Corry to Titusville, a distance of twenty-seven miles, and within the next two years, they extended the line six miles down Oil Creek to Shaffer farm.

The completion of the railroad lessened the necessity for hauling oil long distances over frightful roads; and it decreased the amount of oil run out of Oil Creek by the pond freshets, which meant an increase in toll charges for those who continued to use it. There is good reason to believe that the abandonment of the pond freshet system ultimately came about through the railroad company's gaining control over the dams above Titusville and making it impossible to secure water.[27]

From the beginning, the Oil Creek Railroad had an overwhelming volume of business, and it proved to be one of the best-paying lines in the country. During the first fourteen

[26] *Laws of the General Assembly of the State of Pennsylvania, Passed at the Session of 1860, in the Eighty-fourth Year of Independence* (Harrisburg, 1860), 722.

[27] *Oil City Register*, May 19, 1864.

months, it carried 430,684 barrels of oil, 459,424 empty barrels, 22,727 tons of merchandise, and 59,987 passengers. After deducting all expenditures, the directors declared a dividend of 25 per cent! [28] But this was only the beginning! The Philadelphia & Erie and the Atlantic & Great Western continued to bring passengers and freight to Corry for Titusville faster than they could be moved. The Oil Creek road ran at least two extra trains a day during the spring of 1863, yet it could not handle the business. In December, 1864, the tracks on either side of the stations at Titusville and Shaffer farm were filled with loaded cars, and the warehouses were full of oil.[29] The railroad notified all oil dealers to withhold further shipments until some portion of the track could be cleared. By February, 1865, over 40,000 barrels of oil awaited shipment from Titusville and Shaffer. Such a stagnation of business developed that oil dealers and merchants had to resort to teams and wagons. With only a single track and the demand for cars daily increasing, the situation grew intolerable and stimulated agitation for a new railroad. Despite the inadequate facilities, the Oil Creek Railroad did an excellent business for 1864 and paid a net profit of 53 per cent at the end of the year! [30]

The Atlantic & Great Western Railroad, having captured a large portion of the oil sent to eastern markets from Corry, became interested in gaining control over that being shipped from Shaw's Landing down the canal and from Oil City and Franklin down the Allegheny. In urging the extension of the road from Corry to Meadville and Franklin, William Reynolds, president of the Atlantic & Great Western, wrote on September 6, 1861, "It may from eighteen months' expe-

---

[28] *Crawford Journal*, Jan. 19, 1864. The gross earnings for Feb., 1863, were $24,730.83, and the working expenses were under 25 per cent (*Crawford Journal*, April 7, 1863).

[29] *Crawford Democrat*, Dec. 20, 1864.

[30] *Derrick's Hand-Book*, I, 76. Another writer says that it made 70 per cent profit for 1864 (Titusville *Morning Herald*, July 29, 1865).

rience be safely assumed that this business is permanent and may enter into future estimates of the profits of the road." [31] In view of this and other considerations, it was decided to continue construction. By the fall of 1862 the road had been completed to Meadville, and the next spring a branch line was extended to Franklin. Almost two years elapsed before the Atlantic & Great Western finished its road from Franklin to Oil City. When it was completed in March, 1865, the people of Oil City were truly grateful; gone were the days of struggling through mud to Franklin; and gone was the rapaciousness of the teamsters who charged $2 and $3 for a ride from Oil City to Franklin. They could now ride to Franklin in elegant cars, without mud, in fifteen to twenty minutes for thirty-five cents. The Atlantic & Great Western now encircled the oil region, tapping lower Oil Creek at Franklin and Oil City and the upper portion at Corry. It meant a large and profitable business for the new railroad.

[31] Reynolds, op. cit., 301. This quotation is used by permission of the author.

CHAPTER IX

The Speculative Boom, 1864–1865

COINCIDING with the creation of a market for petroleum at home and abroad and the extension of the railways into the oil region, a wild and unprecedented era of speculation in oil lands and stocks began in 1864 and continued almost to the end of 1865. Almost every farm that had proved valuable for its production, as well as those undeveloped, was purchased at prices far beyond its actual value. Then stock companies were organized for the purpose of developing the property with the hope of making a fortune. The greater demand for petroleum, the magical flowing wells, and the increased confidence in the petroleum business as a permanent thing materially aided in creating the speculative boom, but a number of events occurring in 1864 speeded the movement toward a climax.

One of the most important factors in the development of the speculative boom was the marked improvement in the price of oil. From $3 or $4 at the wells in January, 1864, the price steadily rose to $13.75 in July, the highest since 1860. The enlargement of the market, the decrease in production during 1863, and the advance in the price of gold, beginning back in 1862 and continuing until the summer and fall of 1864, carried prices to a point where oil would pay all expenses and give the owner of a well from $3 to $7 profit. It encouraged the drilling of new wells in old territory and "wildcatting" operations in the new. Moreover, wells which a few months before had seemed worthless on account of the low price of oil now became immensely valuable.

Secondly, the extraordinary career of John W. Steele, more familiarly known as "Coal Oil Johnny," called attention everywhere to the easy money-making possibilities in the oil region. Born on a farm near Oil Creek, Steele became at an early age the adopted son of Culbertson McClintock and his wife, who lived about four miles north of Oil City. When McClintock died, he left the farm to his wife, in trust for Steele. Then came the oil rush. At first, Mrs. McClintock refused to have anything to do with the army of oil seekers; but she finally leased a portion of the farm. Excellent paying wells were struck, and she acquired considerable wealth.

One spring morning in 1864, Mrs. McClintock started to make a fire by pouring crude oil into a warm stove; an explosion occurred, flames enveloped her, and she died from the effects of burns. Upon her death, Steele, who had been working as a teamster, inherited all of the property and accumulated cash. According to legendary accounts, he came into the possession of $150,000; but it is dubious whether the amount exceeded $30,000 or $40,000.[1] Furthermore, it was generally believed that his daily income amounted to $2,000.[2] Whatever the exact amount of cash and daily royalties, this uneducated country youth, without experience in handling money, began a spending orgy that extended over the next twelve months. At once, Steele fell an easy prey to gamblers and unprincipled sharpers of all kinds who robbed and swindled him and encouraged his prodigal and spendthrift ways. Presently, he went to Philadelphia and New York, acquired new companions, became a victim of strong

---

[1] *Crawford Democrat*, Jan. 5, 1867; Titusville *Morning Herald*, Feb. 15, 1868.

[2] James Dodd Henry, *History and Romance of the Petroleum Industry*, 290; *Crawford Democrat*, Jan. 5, 1867.

drink and bad company, and proceeded to squander his fortune. Because he derived his wealth from oil, newspapers nicknamed him "Coal Oil Johnny" and published the most sensational stories about his escapades.

One of the most widely heralded stories, which gave him the greatest notoriety, related to his acquisition of a minstrel troupe. While in Philadelphia, Steele saw a performance of Skiff and Gaylord's minstrels, became greatly excited over the performance, and bought a third interest in the show with John W. Gaylord as a partner.[3] Steele proceeded to give each member of the cast a diamond ring, a gold watch and chain, together with a complete wardrobe, and then he took the show on tour, doing a large business everywhere. Reaching Utica, New York, he treated the company to a supper which cost $1,000. Steele then conceived the idea of traveling on a private train; so they canceled all dates for two weeks and the troupe went junketing with Steele paying the bills. Their dates having been canceled, Steele insisted on indemnifying the company for the loss of time and paid all salaries, computed the probable receipts based upon packed houses, and paid over that sum to Seth Slocum, the treasurer.

In Chicago, Steele leased the Academy of Music, where they did a big business, and at the end of the season he proposed to have a benefit for Skiff and Gaylord and sent a representative to engage the Crosby Opera House. The manager insolently replied, "We won't rent our house for a —— nigger show!" Excited and angry, Steele went to the owner's office and pulling a roll of bills from his valise, counted out $200,000 and offered to buy the Opera House. Thunderstruck, the owner said that if Steele was that sort of man, he could have the Opera House free. The benefit

[3] "Coal Oil Johnny," *Petroleum Age*, VII (1888), 16; *Titusville Morning Herald*, Dec. 10, 1870.

was held, and the next day Steele found one of the finest carriage horses in Chicago and presented it as a gift to the owner.

Other stories freely circulated about Steele's renting the Continental Hotel in Philadelphia for one day for $8,000, buying the Girard House in the same city, lighting his cigars with $100 bills, tipping bootblacks $50 or $100, and giving his male and female friends expensive diamonds. A large percentage of them were probably without foundation; others were exaggerated and distorted; and some of them were true.[4] That Steele freely, lavishly, and rapidly spent his money in riotous living is beyond question. However, instead of spending millions, as is commonly believed, it has been conservatively estimated that $500,000 would cover the squandered fortune.[5]

With his money gone, his professed friends deserted him, and Steele took a job as doorkeeper for the minstrel troupe in which he once had owned an interest. Within a short time he returned to Rouseville, got a job as baggage agent on the Oil Creek Railroad, and tried to earn a living and forget his past. His former creditors quickly caught up with him, for in 1866 some Philadelphia lawyers, armed with a judgment for $70,000 in favor of a Philadelphia hotel, jeweler, and tailor, caused the sheriff to levy against Steele's farm; but the United States prevented its sale on account of a prior lien for taxes.[6] A few weeks later the local sheriff sold the farm for $35,000 subject to the claim of the United States. Finally, in February, 1868, Steele filed a voluntary petition in bankruptcy with the United States District Court in Pittsburgh. His indebtedness amounted to over $100,000 and, among other items, included the following:[7]

[4] John Washington Steele, *Coal Oil Johnny* (Franklin, Pa., 1902), 114.
[5] James Dodd Henry, *History and Romance of the Petroleum Industry*, 290.
[6] Titusville *Morning Herald*, Jan. 5, 1867.
[7] *Ibid.*, Feb. 15, 1868.

Henry W. Kanaga of the Girard House, Philadelphia $19,824
Wm. A. Galbraith, lawyer, Erie 10,000
J. E. Caldwell & Co., Philadelphia, jewelry 5,805
John D. Jones, harness 1,250
Wm. Horn & Co., cigars 562
E. H. Conklin, Philadelphia, liquor 2,024
Phelan & Collender, Philadelphia, billiard tables . 1,500
Unknown creditor, oil paintings 2,200
For hats 300

The wide publicity about his early career and subsequent misfortune soon led Steele to abandon his position at Rouseville, move west to Iowa and later to Nebraska, where he raised a family and gave his children a good education.

The publicity relating to the huge profits derived by a few of the more successful oil companies acted as a powerful stimulant to the speculative movement. During the last six months of 1863, the Columbia Oil Company paid its stockholders over $300,000, in a dividend amounting to more than $26 a share.[8] On account of the extraordinary dividend, the stock of the company rapidly appreciated, and it was soon worth considerably more than par value on the market. To correct this situation, the capital stock was increased from $200,000 at $20 a share to $2,500,000 at $50 a share. From its earnings for July, 1864, the company divided $100,000 among the stockholders and had 10,000 barrels of oil on hand, worth over $100,000; so it had made over $200,000 in one month![9] For the same month, the Densmore Company declared a dividend of 5 per cent, and the Noble and Delamater Petroleum Company declared its fourth monthly dividend of 10 per cent.[10] If these companies could do this, it was easy to believe that others might be as successful.

Lastly, the opening up of new territory on Cherry Run during the summer of 1864 further intensified the excitement.

[8] *Crawford Journal*, Dec. 15, 1863.
[9] *Oil City Register*, Aug. 18, 1864.
[10] *Derrick's Hand-Book*, I, 41.

Little or no effort had been made prior to 1864 to extend drilling operations into the valleys formed by streams empty-ing into Oil Creek and the Allegheny River. With so many productive wells on the Creek and the supply exceeding the demand, operators did not need to seek any other locality. Furthermore, the early oilmen had limited means and were reluctant to invest in lands other than those which had been tested and promised an immediate return. For these reasons, no extensive developments had taken place elsewhere.

During the summer of 1864, however, William Reed started drilling on the Rynd farm two miles up Cherry Run, which flowed into Oil Creek from the east about three miles above Oil City. The valley had been almost entirely neglected, and land was comparatively cheap. A mere mention of the possibility of finding oil on Cherry Run before 1864 caused a smile from most of the operators; a majority of them looked upon it as "dry diggin's." In fact, Reed had offered to sell his property for $1,500 in 1863, but no purchaser could be found.[11] Before completing his well, he was joined by R. Criswell and I. N. Frazier, who had a little ready cash. When down to the proper depth, the well had the appearance of a "duster"; but after several days of pumping and testing, it began flowing on July 18, at the rate of 280 to 300 barrels.

Criswell, after realizing $30,000, sold his one-fourth interest in the land for the princely sum of $280,000; Reed made $75,000 from the sale of oil, then sold his one-half interest for $200,000; Frazier, who owned one-fourth of the property, received more than $100,000 from the sale of oil and then disposed of his interest for $100,000.[12] Thus the original proprietors made a total of $785,000 from the Reed well within ninety days. Three other wells were soon drilled on the same lease, and the four produced steadily for the next

[11] Bone, *Petroleum and Petroleum Wells*, 85.
[12] J. T. Henry, *The Early and Later History of Petroleum, etc.*, 230–231.

two or three years. The new owners also made at least
$785,000 from the purchase, and all together it is estimated
that the profits of this one oil transaction approximated
$2,000,000.[13]

The Reed well opened up a new and untested area and
precipitated a mad scramble for land along the entire length
of Cherry Run. The rush was tremendous, the excitement
great, and land once thought worthless sold for fabulous
sums. The Smith farm of fifty acres, above the Reed well,
had been offered for sale in '62 and '63 for $250 over the
encumbrances, but now the owners refused an offer of
$4,000,000.[14] Just above the Smith farm lay 400 acres, which
embraced the entire valley for a mile and a half. C. Curtis
purchased the farm in 1862, at which time the investment
seemed anything but judicious; in 1864 the property constituted the most valuable of any in the oil region. Curtis
now organized the Cherry Run Petroleum Company with a
capital stock of $200,000. Though at first it drilled its own
wells, the company's subsequent decision to sell a few leases
created fierce excitement not only locally but in New York,
where its stock with a par value of $10 suddenly shot up
from $7 a share to $32. This was a breath-taking affair for
many persons who, only a short time previously, had said
that anyone investing on Cherry Run was ready for the in-
sane asylum.[15] The fact that the company had hardly started
to develop the property and had never declared any dividends
made it all the more amazing.

Late in August all the land on Cherry Run that could be
leased had been taken up, and the contiguous territory for
several miles had been purchased or leased. Derricks soon
covered the whole valley and the steep hillsides. Engines and
tools lay about in all directions, and a greater degree of ac-

13 *Ibid.*
14 *Derrick's Hand-Book,* I, 43.
15 *Oil City Register,* Nov. 24, 1864.

Mather Photograph

FRAZIER WELL, HOLMDEN FARM, PITHOLE, IN MAY, 1865

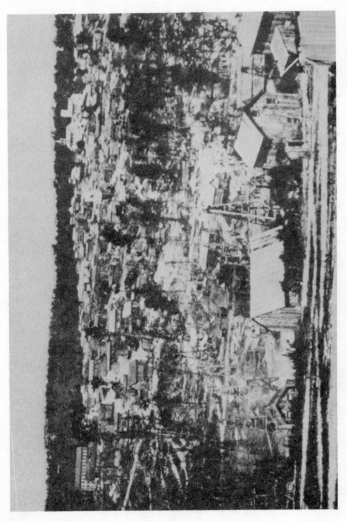

Mather Photograph

PITHOLE, 1865

tivity prevailed here than at any other point in the oil region. By November, Cherry Run had become the favorite place for oil seekers. Not a single well had failed to produce oil, and the daily yield amounted to about 1,000 barrels.

One result of all these factors working jointly was that speculation in land increased tremendously and land values in many cases doubled and trebled. Along Oil Creek during 1864 the King farm brought $85,000; the Parker, $100,000; the Siverly, $100,000; the Funk, $150,000; the Hyde and Egbert, $200,000; the Crocker well and four acres on the Foster farm, $220,000; and the Noble and Delamater well and territory $300,000.[16] Tarr was offered $800,000 in cash for his farm, but refused.[17] The sale of the Blood farm in April, 1864, for $650,000 in greenbacks eclipsed all real estate transactions; it was the largest amount of cash that had ever been paid for any oil land.[18] On it were some thirty producing wells, yielding from 400 to 700 barrels a day. The daily cash income was estimated at $2,400. With the exception of three or four tracts, the Blood farm was the most productive in the whole region. This transaction held the record for only a few months, however, as the Graff, Hasson farm at Oil City was sold in November for $750,000.[19]

The success of operators on Cherry Run led speculators to buy up hitherto partially developed or wholly neglected lands along the tributaries of Oil Creek and on the Allegheny. Typical of what happened is the case of a farm on the east branch of Cherry Run, which sold for $7,000 in the fall of 1864, $15,000 on January 14, 1865, $33,000 two days later, and $500,000 on February 20, 1865.[20] Every spot of ground anywhere, having the slightest appearance of being oil land,

[16] Derrick's Hand-Book, I, 37, 39; Crawford Journal, April 12, 1864.
[17] Titusville Morning Herald, Oct. 30, 1865.
[18] Warren Mail, June 4, 1864; Venango Spectator, April 27, 1864.
[19] Oil City Register, Nov. 24, 1864.
[20] Amasa M. Eaton, "A Visit to the Oil Regions of Pennsylvania," Western Pennsylvania Historical Magazine, XVIII (1935), 191.

for a space probably not less than 100 miles square in the aggregate was either purchased or negotiated for.[21]

Another result of these different factors at work was the formation of an unprecedented number of stock companies for the purpose of buying or leasing land and drilling wells. In September, 1864, the total number in existence exceeded 100 with a nominal capital of over $52,000,000; by February, 1865, there were over 500, representing an aggregate capital of $356,565,000.[22] So many companies had been organized in New York that it was almost impossible to obtain suitable office space in the financial district, and the number daily multiplied. During the last week in March, over twenty were formed in eastern cities, representing a capitalization of $12,500,000. In view of the unusual situation which had developed, the *New York World* declared, "The inducements to permanent investors are unparalleled." [23]

Irrespective of the character of the company, thousands of desperately excited citizens purchased stock, believing that they now had a new road to wealth. The swiftness with which it was subscribed seemed perfectly marvelous: several companies were reported as having sold out in a single day, and one is said to have sold every share in four hours.[24] The daily sales of petroleum stocks at the regular stock exchange of Philadelphia in the fall of 1864 amounted to more than $200,000.[25] Thirty, forty, fifty, and sixty thousand shares of stock changed hands daily. According to the *Philadelphia Ledger,* there was a "lively demand for oil stocks, and the shares of new companies, which as yet have not a single well in operation, are selling at a large advance on the original cost. This is accounted for as well by the plethora of money seeking investments as by the immense profits being realized

21 *Oil City Register,* Aug. 18, 1864.
22 *Crawford Journal,* Sept. 13, 1864; *Venango Spectator,* Feb. 22, 1865.
23 Quoted in the *Venango Spectator,* Jan. 25, 1865.
24 *Venango Spectator,* Jan. 25, 1865.
25 *Derrick's Hand-Book,* I, 42.

by those companies in successful operation."[26] So great was the demand for stocks in New York that a Petroleum Board was organized in the fall of 1864. Other cities, especially Pittsburgh, Baltimore, Boston, and Cincinnati, had experiences similar to those of Philadelphia and New York. The favorite shares, observed Sir S. Morton Peto, the London banker, were not £25, £20, or even £10 a share as they would be with Englishmen, but shares from two to four shillings each. According to Peto there were "hundreds of thousands of provident working men, who prefer the profits of petroleum to the small rates of interest afforded by savings banks."[27] The amount of capital thus withdrawn from other pursuits or savings and applied to the purchase and development of oil territory cannot be accurately estimated, but one authority claims it exceeded $100,000,000.[28]

The speculative fever was not confined to the ordinary citizens; it affected some of those occupying high positions in the financial world. During the summer of 1864 William Moorhead, one of the partners of Jay Cooke and Company in the Washington office, organized an oil company, the stock of which was soon to be offered to the public at $7 a share.[29] In his opinion, it was so good that he invested the money of widows and children in it and offered some to his partner, Harris C. Fahnestock, for $5 a share. Fahnestock took 1,000 shares for himself, 500 for William S. Huntington, cashier of the First National Bank of Washington, and 1,000 for the oldest and best clerks in Cooke's Washington office. Later, an agent of an oil company promoted by Simon Cam-

[26] Quoted in the *Crawford Journal*, Aug. 16, 1864.

[27] S. Morton Peto, *The Resources and Prospects of America, Ascertained During a Visit to the United States in the Autumn of 1865* (London, 1866), 204.

[28] Hayes, "Report of the United States Revenue Commission on Petroleum as a Source of National Revenue," *House Exec. Doc.* No. 51, 39 Cong., 1 sess., 7.

[29] Henrietta M. Larson, *Jay Cooke* (Cambridge, 1936), 153; Ellis Paxson Oberholtzer, *Jay Cooke: Financier of the Civil War* (Philadelphia, 1907), I, 439–441.

eron of Pennsylvania came to the Washington office with a request that Henry Cooke and Fahnestock serve as directors. Although they were to be given 1,000 shares free for the use of their names, they declined. A few days afterwards the agent returned to say that Cameron had assumed the responsibility for placing the name of Henry Cooke beside his own on the list of directors. It was soon published in the advertisements, and when Jay Cooke discovered the fact he vigorously protested against such "wildcat operations" and brought the first test of his headship of the firm. "Not one out of twenty of the 'respectable' oil companies," Cooke told his partners, "are really worth the paper they are printed on. If I am not to know what ventures my junior partners propose to make, what means they propose to withdraw from the firm or borrow outside (which is the same thing), or in other words what departures are proposed from the strict and legitimate course of business, why, I wish to know it. It is my right as I understand the ordinary nature of such a partnership as ours." [30] Henry Cooke, therefore, told Cameron that his name would have to be removed from the board of directors, thus closing the incident. The oil transaction of Fahnestock continued to annoy Jay Cooke, and finally the former sold all of his shares.

Also caught in the whirlpool of speculation were James A. Garfield and other members of Congress. A former staff officer, Captain Ralph Plumb, suggested to Garfield in January, 1865, that he buy or lease some oil land on the Big Sandy. Garfield significantly replied, "I have conversed on the general question of oil with a number of members who are in the business, for you must know the fever has assailed Congress in no mild form." [31] He suggested that Plumb make an

[30] Larson, *op. cit.*, 153. This quotation is used by permission of the President and Fellows of Harvard College.

[31] Theodore Clarke Smith, *The Life and Letters of James Abram Garfield* (New Haven, 1925), II, 822. This and the subsequent quotations are used by permission of the author and the Yale University Press.

examination and report at once. A few months later, Plumb received another letter about the matter from Garfield, who said, "It would aid us greatly if we could be able to make a full representation of the prospects to leading men here before Congress adjourns." [32] The report must have been favorable, for Garfield, Plumb, and others organized a company and purchased over 4,000 acres of oil land.[33]

In the oil towns, the full effects of the speculative boom were felt. At Corry, hundreds of people waiting to catch the train for Titusville could not secure a night's lodging, and the miserable little depot of rough boards was utterly unable to shelter one-half the crowd.[34] One of Jay Cooke's agents wrote from Titusville in March, 1865: "The people through this country can't talk connectedly about anything but oil. They tackle me with oil talk and I give them seven-thirty, but they can't understand it. The bankers are all the same. No matter what you say to them it winds around to oil." [35] At Oil City, the editor of the *Register* urged the immediate construction of 100 or more boarding houses to feed the crowd.[36] Of the effect upon Warren, Pennsylvania, the editor of the *Warren Mail* wrote: "The oil and land excitement in this section has already become a sort of epidemic. It embraces all classes and ages and conditions of men. They neither talk, nor look, nor act as they did six months ago. Land, leases, contracts, refusals, deeds, agreements, interests, and all that

[32] *Ibid.*

[33] *Ibid.* Garfield wrote on May 11, 1875, "In 1865 I made a successful venture in the purchase and sale of oil lands from which I realized about six thousand dollars" (*ibid.*). Garfield also had another connection with oil. As an attorney, he assisted the Phillips brothers of Newcastle, Pa., to organize an oil company in Chicago. While engaged in this work, he said, "It is a work almost as great as a military campaign, but if we succeed it will go far toward settling one of the most important and difficult problems of my life which relates to our living" (*ibid.*, 823).

[34] Oberholtzer, *op. cit.*, 615; Bone, *op. cit.*, 56.

[35] Oberholtzer, *op. cit.*, 615. This quotation is used by permission of the Macrae Smith Company, Philadelphia, Pa.

[36] *Oil City Register*, March 2, 1865.

sort of talk is all they can comprehend. Strange faces meet us at every turn, and half our inhabitants can be more readily found in New York or Philadelphia than at home in Warren. The court is at a standstill; the bar is demoralized; the social circle is broken; the sanctuary is forsaken; and all our habits, and notions, and associations of half a century are turned topsy-turvy in the headlong rush for riches. Some poor men become rich; some rich men become richer; some poor men and some rich men lose all they invest. So we go." [37] The editor prophetically ended by declaring, "The big bubble will burst sooner or later."

[37] *Warren Mail,* March 18, 1865.

CHAPTER X

Pithole

DESTINED to be the scene of the greatest speculation in land was the territory along Pithole Creek, a small stream flowing southward and emptying into the Allegheny River about eight miles above Oil City. This creek received its name from the fact that an upheaval of rock had left a series of deep pits on the hills near its mouth; from the largest of these a current of warm air repelled leaves or pieces of paper, and snow quickly melted around the cavity. The valley had attracted the attention of oil hunters almost from the time of Drake's discovery, but the low price of oil, the cost of experimenting, and the excitement along Oil Creek prevented any extensive operations from being undertaken.

Lying on both sides of Pithole Creek, about four miles from Plumer and six miles from the Allegheny, was the 180-acre farm of Thomas Holmden; to the west was the Walter Holmden farm, to the north the Copeland, to the east the Hyner, and to the south the Rooker. Living here in log cabins in a dense forest, isolated and inaccessible, the Copelands, Rookers, and Holmdens spent their time raising buckwheat and hunting deer. In the spring of 1864, I. N. Frazier and James Faulkner, employees of the Humboldt refinery at Plumer, and others organized the United States Petroleum Company and then leased sixty-five acres of bottom-land on the Holmden farm for twenty years, giving one-fourth the oil.[1] Selecting a wild and secluded spot by means of the witch-

[1] *Oil City Register,* June 1, 1865.

hazel twig, the company commenced to drill. It was purely a speculative venture, and well informed oilmen regarded it as chimerical. Before the completion of the well, however, Thomas G. Duncan and George C. Prather purchased the Holmden farm, subject to the lease, for $25,000. They at once offered the farm for sale in Philadelphia, but on January 7, 1865, the well began flowing 250 barrels. The stock of the United States Company rocketed from $6.25 to $40 a share, and the owners of the farm immediately withdrew the property from the market.[2] Owing to the manner in which the contract had been drawn, Holmden possibly could have recovered his property, but Duncan and Prather finally agreed to pay $100,000, instead of the $25,000, and give him the royalty of the well for one week.[3]

For a few weeks the Frazier well basked in all its glory, and then, early in April, a Boston company completed a well, known as the Homestead, on the Hyner farm scarcely 100 feet from the Holmden line; it pumped fifty barrels. One of the owners had been John Wilkes Booth, who paid $13,000 for a thirteenth interest in the fall of 1864; but, after retaining it only a few months, he sold out to one of his partners. On the night that Booth assassinated Lincoln, according to the story, a visitor held a lighted cigar too close to the mouth of the oil pipe, setting fire to the well. After the blaze had been extinguished, the well increased its flow to 250 barrels.

On account of the discovery of oil on land hitherto regarded as unpromising, in such great quantities and at a considerably higher altitude than Oil Creek, the Frazier and the Homestead wells created a tremendous demand for land, stimulated the drilling of new wells, and precipitated a general stampede to Pithole Creek, which soon exceeded anything ever seen in the oil region. The Holmden farm be-

[2] *Titusville Morning Herald*, April 5, 1871; *Venango Spectator*, Jan. 25, 1865.

[3] *Oil City Register*, June 1, 1865; Titusville *Morning Herald*, May 11, 1866.

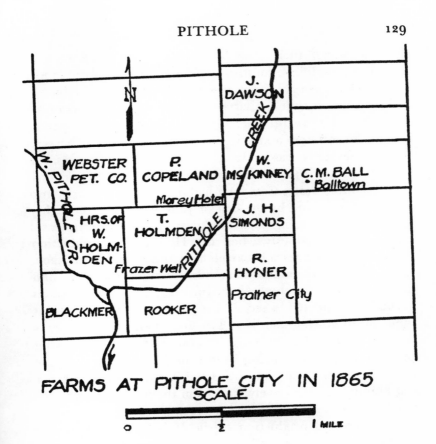

FARMS AT PITHOLE CITY IN 1865

SCALE

0   ½   1 MILE

came the Mecca for everyone interested in oil and anxious to make money quickly. They came by the hundreds. Every road approaching Plumer and that between Plumer and Pithole Creek were filled with persons on foot, on horseback, or in stages and wagons; and on arrival they rushed breathlessly up and down the creek with their pockets full of greenbacks, eager to buy leases or land at any price.

The rush and excitement was intensified by several other influences which simultaneously conspired to foster its growth. The Civil War had just closed, bringing peace and leaving the country flooded with an inflated currency; capitalists were eager to invest their greenbacks and make more money; thousands of soldiers, constituting a potential supply of labor, had been discharged from the army and waited to be absorbed in peace-time activities; and the speculative craze of '64 and '65 was at full tide, with hundreds of oil companies hungry for territory and ready to lease or buy land wherever a promising strike was made. Under these circumstances, the oil bubble on Pithole Creek assumed enormous proportions.

The United States Company had divided its property into half-acre leases, and by the first of June had sold over eighty for a bonus, averaging $6,000 an acre.[4] Speculation in leases increased tremendously when the Homestead's production suddenly jumped to 500 barrels and the Frazier's to 1,200. Leases on the Holmden farm now doubled in price, and the excitement ran high. On June 17, the United States Company completed its second well, below the Frazier, and it commenced flowing 200 barrels at first, then 800. Two days later, a third well just above the Frazier started off, flowing 300 to 400 barrels. Having been completed about the same time, they came to be known as the Twin wells Number 1 and Number 2 respectively, and from them the company derived a daily income of at least $2,000.

The increased production of the Homestead and Frazier and the successful completion of the Twins created more confidence in the territory, and owners of land found it easy to sell leases on almost any terms. Under the shade trees around the celebrated Frazier well, an excited crowd talked and bargained, and a large amount of outdoor cash business was conducted here. Fabulous prices were paid for land which for any other purpose was worth nothing. Leases between the Frazier and the Homestead wells sold for $5,000 to $10,000, and by the end of June the United States Company had derived over $100,000 from the sale of leases.[5]

[4] *Oil City Register,* June 1, 1865.
[5] *Ibid.,* June 22, 1865.

Meanwhile, speculators overran the 100-acre farm of Jesse Rooker, which lay on both sides of the creek. Any man offering $2,000 for it prior to 1865 would have been considered insane. With the completion of the Frazier well, however, the speculators visited with Rooker and frequently offered to buy the farm; to all of them he turned a deaf ear. Though ignorant of oil beneath the surface, Rooker was obstinate and superstitious enough to believe that his farm was the most valuable on earth. After much persuasion, however, he finally sold his farm in June to J. W. Bonta, James A. Bates, and others for $280,000; these gentlemen, in turn, divided the farm into 180 half-acre leases and began selling them at bonuses ranging from $3,500 to $7,500.[6]

In the midst of all the bargaining and drilling of wells, the unexpected bursting of two or three storage tanks full of oil at the Frazier well on June 19 created a panic. Twenty-five hundred barrels of oil went rushing down Pithole Creek. Less than a half-mile below, two boys applied a match to the stream, which ignited and the raging fire quickly spread up the Creek, menacing the Frazier well. When the smoke was observed in Plumer, everyone felt certain that the Frazier had caught fire; anxious stockholders rushed to the telegraph office to send word to their Wall Street brokers saying, "Sell my United States stock at any price." By exerting great effort, over 200 men finally extinguished the fire. An agent of the United States Company then rode to Plumer and announced that the great well was now out of danger. Nervous stockholders, who had just set Wall Street in a flurry, took new courage and wired their brokers, "Buy all the United States you can get."

At the end of June, the four flowing wells on Pithole Creek were producing over 2,000 barrels daily, or one-third of the total production for the whole oil region, and the prospects of a greater yield were bright, for scores of wells were being

[6] Titusville *Morning Herald*, Aug. 24, 1865.

drilled.[7] The Island well, a few rods above the Homestead, added to the daily production when it commenced flowing 250 barrels, jumping later to nearly 900. About the same time, Twin Number 2 increased its yield to nearly 1,000 barrels; but the Frazier fell off to about 700, and many believed that it had seen its palmiest days.

At least 3,000 teamsters were engaged in hauling oil to the Allegheny River, Titusville, Shaffer farm, Miller farm, and Oil City.[8] Almost any time of the day, one could see an endless procession of empty and loaded wagons, sometimes three or four abreast, gorging the narrow roads, horses tugging and snorting, drivers lashing and swearing, making their way to the various shipping points. The roads were none too good in dry weather, but in wet they were wretched. They were so impassable late in July that teamsters were unwilling to go with a load, and teaming almost ceased. To fill up the mudholes and help keep the road in repair, oil shippers and merchants at Titusville raised a maintenance fund, and the teamsters contributed $1 a week to the cause. With horses and wagons scarce and roads bad, teamsters charged $3 a barrel for hauling, which was as much as the price of a barrel of oil; and then they threatened to strike for higher pay.

Typical of the way in which speculators made money was the case of T. S. Sumner and F. M. Pratt of Tarr farm, who purchased three acres of the Rooker farm in July for $75,000 before any oil had been struck there. They in turn sold leases on the three acres and interests in wells amounting to $175,000, and they still had a few leases left.[9] One of the persons purchasing from Sumner and Pratt sold half of his interest for $7,000 more than he gave for the whole interest!

Beginning in July a speculative rage in one-sixteenth interests in the wells started and such an interest in the Frazier,

[7] *Oil City Register*, June 29, 1865.
[8] *Business Directory of Pithole City 1865–66*, 6.
[9] Titusville *Morning Herald*, Oct. 24, 1865; *Oil City Register*, Feb. 8, 1866.

the Twins, or the Homestead easily sold for $1,000 to $3,000.[10] For the benefit of those who could not buy an interest on account of the exorbitant prices, a group of men organized the "Working Men's Pithole Creek Oil Association" for the purpose of buying sixteenth interests in different wells being drilled near the Frazier.[11] The shares sold for $10 apiece and entitled the purchaser to a proportionate interest in all wells completed after the payment of his subscription. With the funds thus distributed among several wells, the stockholders had greater chances. The stock sold rapidly in amounts ranging from one to twenty shares.

The favorite prey for unscrupulous speculators was a person with a bulging pocketbook or an agent of an eastern oil company, and all sorts of tempting bait were used to get him to buy. Frequently the wells were "doctored" by pouring in barrels of oil and dumping buckets of fresh sand on the floor at night. When the prospective buyer arrived the next morning, all evidence pointed to a successful well. In several instances, wells, sold at fabulous prices, had been connected by means of underground pipes with tanks of oil at a distance. When the "suckers" came to test the well, the subterranean pipe was opened, oil ran into the tubing, and it was pumped up as if coming directly from the sand. As soon as the "suckers" paid over the cash, the well died of thirst.

Eclipsing all speculative transactions was the sale of the Holmden farm in July to George J. Sherman, Henry E. Pickett, and Brian Philpot of Titusville for $1,300,000![12] It involved the largest sum ever paid in the oil region for any single tract of land. Sherman at once started looking for parties to whom the contract could be sold. Since the first payment of $400,000 was due at the end of thirty days and Sherman did not have a buyer, the contract was renewed and

[10] Titusville *Morning Herald*, July 14, 1865.
[11] *Ibid.*, Aug. 14 and 26, 1865; *Venango Spectator*, Aug. 23, 1865.
[12] J. T. Henry, *The Early and Later History of Petroleum, etc.*, 237–239.

extended two weeks. Finally, aided by three Chicago men, Sherman obtained $400,000 in greenbacks, and as the contract called for the first payment to be made on the farm, the four left for Titusville. Reaching that place, each gentleman armed himself with a pistol, took one-fourth of the money, and started overland for Pithole Creek. On the last day of grace and in the late afternoon, these men safely reached their destination, went directly to Prather and Duncan's office, and laid the money on the counter. Claiming that the life of the contract expired with the setting of the sun, Prather and Duncan declined the $400,000. Subsequently, a suit was started in the United States District Court at Pittsburgh, but the matter was finally adjusted without going to trial.

Two of the largest wells were completed on the Holmden farm in August, intensifying the labor and ambitions of everyone. The Grant well, named in honor of the hero of Appomattox, started flowing on August 2 at the rate of 200 barrels. The next morning the oil flowed faster than ever, and it soon spouted 1,200 barrels, becoming the prince of wells. The working interest in this well was widely scattered, no person holding over a thirty-second interest. The lot adjoining the Grant well had been purchased in the spring for $1,600 and now that it was regarded as the choicest lease on the market, it sold for $16,500, the highest price paid for any half-acre lease.[13] The second well, a short distance from the Frazier, commenced on August 28 to shoot oil sixty feet in the air and gas ten to fifteen feet higher; it yielded 300 barrels at first, then jumped to 1,500. The Pool well, as it was called, now became the largest producer at Pithole.

An increasing amount of drilling took place on both the Rooker and the Morey farm in August. The first well on the Rooker farm was completed and started flowing late in August at the rate of 400 barrels, and it steadily improved. A twenty-fourth interest in the land, which had been selling heretofore

[13] *Oil City Register*, Aug. 24, 1865.

at $25,000, now brought $30,000 to $40,000.[14] Although it pumped only forty barrels daily, the Hoosier well on the Dawson farm, about a quarter of a mile above the Island and Homestead wells, created considerable interest in the surrounding territory. On account of it, the Morey farm especially became a favorite place; it was literally covered with derricks, and more wells were being drilled there than on the Holmden.

The excitement on Pithole Creek reached its height in September, when the daily production amounted to 6,000 barrels, which was two-thirds of all the oil produced in the oil region.[15] Without a rival, the Holmden farm had ninety-six wells either producing or being drilled, and a daily production of almost 4,000 barrels. As a climax to the reckless speculation of the past few months, Prather and Duncan now sold the Holmden farm for $2,000,000! [16]

While all these events had been happening, Duncan and Prather laid out a town on the Holmden farm in May; and without any ceremony or official designation it came to be called Pithole. There were 500 lots, each with a frontage on the principal thoroughfare, Holmden Street, which ran northwest for a half-mile. Beginning near the United States well they laid off First Street and others at right angles to it. Strange to say, the lots could not be purchased: they were leased for three years with the privilege of removing the buildings on the expiration of the lease or selling them to the owners; or they were leased for five years, at the end of which the lessee left the buildings and improvements without compensation other than two years' extension of time.[17] Under these arrangements, most of the city would revert to the owners of the land in three years' time and the rest in five. Despite the peculiar leasing system, most of the lots were leased within

[14] *Ibid.*
[15] *Derrick's Hand-Book*, I, 51.
[16] Titusville *Morning Herald*, Sept. 21, 1865.
[17] *Pithole Daily Record*, Oct. 5, 1865; *Oil City Register*, June 1, 1865.

a week at the rate of $275 a year. By July, however, the demand was so great that leases were selling for $850 each and advancing at the rate of $50 a day.[18]

On May 24 there was but one building under construction; a week later nearly the whole of Holmden Street was solidly lined with buildings, and hundreds of others were under construction.[19] The tall pine, soft maple, and old oak trees rapidly disappeared in the rush to build. Everywhere the sound of pounding hammers and the heavy crashing of trees could be heard. On account of the leasing scheme, tenants simply put up one- and two-story shanties, without lath or plaster, and they could not be erected half as fast as needed. Frequently, men entered into contracts to build a two-story dwelling or business building and have it completed and ready for occupancy within five days after signing the contract. It is not surprising, therefore, that the structures appeared like a sieve when illuminated at night or that more than one fell down before entirely completed. On account of the haste with which they were erected, a correspondent of the *Nation* referred to Pithole as "a gigantic city of shreds and patches."

Those who came to Pithole were forced to endure all sorts of personal hardships and privations. Many persons profited by hauling drinking water from distant wells and by retailing it for ten cents a drink, one dollar a bucket. Under the circumstances, the liquor business flourished. "Whiskey abounds," declared one visitor. "Every establishment sells the beverage. Taken in any quantity it is sickness, and if indulged in to any extent, certain death."[20] It was even difficult to get anything to eat, as food could not be imported fast enough. When one could get something, it was expensive and bad. "Tough beef, bread, and a decoction of unknown ingredients called coffee," wrote a correspondent of the *Erie*

18 Titusville *Morning Herald,* July 3, 1865.
19 *Oil City Register,* June 1, 1865.
20 Titusville *Morning Herald,* July 29, 1865.

HOLMDEN STREET, PITHOLE, AUGUST, 1895

MILLER FARM STATION, OIL CREEK, 1865

Mather Photograph

Observer, "can be secured at the moderate sum of $4 per day, or from $8 to $15 per week." [21] Another traveler said, "Judging from the impossibility of mastication, I would say that by some mistake the carcass had been sent to the tannery and the hide served up at our hotel." [22] But who cared if the beef was tough, bread nine days old, and coffee without cream? Everyone intended to make millions and then return to the "States." At night they slept in piles of shavings, haylofts, and on pine boards; fifty to seventy-five slept nightly in a barn for which they paid $1; and many slept in tents or outdoors on the ground. A chance to sleep under any kind of shelter was considered a luxury.

Without any sewage disposal system, some of the people carried their garbage into the woods and dumped it; large boarding houses and hotels simply threw it out the back door. One visitor reported, "The whole place smells like a camp of soldiers when they have the diarrhoea." [23] With board bad, provisions scarce, sultry weather, a lack of sanitary facilities and good drinking water, several cases of fever and dysentery broke out. Rumors immediately spread everywhere that Pithole had a fever epidemic, even cholera, which temporarily deterred many people from coming and acted as a brake on business.

As the people poured into Pithole during the summer and the town developed, some of the conveniences, comforts, and institutions of settled communities began to appear. In July, the government opened a post office which handled over 5,000 letters a day. So great was the volume of business that the postmaster and clerk were both on duty from seven A.M. to two A.M. Since there were few boxes, the mail had to be sorted and distributed from the general delivery window. Men and boys, therefore, pushed and crowded trying to get

[21] *Ibid.*
[22] *Ibid.*, Aug. 15, 1865.
[23] *Ibid.*, July 29, 1865.

to the window, and frequently sold out their places to impatient letter-seekers for $1 and up.

Some of the substantial elements undertook to provide for religious worship, and the first preaching services were held in July in an unfurnished hotel by a United Presbyterian and a Methodist minister. These two at once started to raise money for erecting churches for their respective denominations and the liberality with which people contributed indicated that material prosperity wasn't everything at Pithole. Prather and Duncan donated two of their best lots and also made a generous subscription to each building fund. Sufficient money had been given by mid-August to proceed with the construction of the two churches, each capable of seating 500 people and costing about $9,500.

For those who craved entertainment and pleasure, W. H. Murphy and brother constructed a theater on First Street with a seating capacity of 1,100. The interior was handsomely painted and decorated, and chandeliers from Tiffany's furnished the light. A first-class dramatic company opened the theater on September 17, and played every night thereafter, except Sundays. One night after "Macbeth," the leading lady responded to a curtain call, and the appreciation of the audience was demonstrated by tossing $500 onto the stage.

Some of the leading citizens organized the Swordsman's Club for social purposes. They elegantly furnished some rooms, and good-fellowship prevailed at all their meetings. The motto of the club was "R. C. T." meaning rum, cards, and tobacco. When some jokester told a minister that it meant "Religious Counsels Treasured," it is said that he preached a sermon eulogizing the organization. The club gave elaborate concerts and balls, to which all the best people of oildom were invited, and which were the outstanding social events at Pithole.

No less notable was the organization Pithole's Forty Thieves, which was composed of well superintendents of outside companies. Unable to understand the difficulties in drilling wells, distant stockholders blamed superintendents for the lack of dividends and often called them "slick rascals," "plunderers," and "robbers." Some practical jokers suggested that the superintendents ought to organize a club, the idea took, and Pithole's Forty Thieves came into existence. Social meetings were regularly held, and the favorite pastime was teasing innocent persons who really believed that stealing was the object of the organization.

In spite of the great influx of all sorts of people and the excitement, good order and respect for individual rights generally prevailed.[24] Often it was the subject of remark and surprise among strangers that they seldom saw any drunken or disorderly conduct at Pithole. Toward fall, however, disorder gradually increased, street fights became common, drunkenness more prevalent, garroting almost common, and at least one robbery a day was committed.[25] In the absence of any police, most citizens resorted to carrying pistols in case they were attacked.

By September, Pithole had two banks, two telegraph offices with a business larger than those of Titusville and Oil City, a daily newspaper, a waterworks system, a fire company, two churches, scores of boarding houses, grocery stores, machine shops, and other businesses. There were over fifty hotels, and some, like the Morey, Chase, and Bonta, were large, elegant, and comfortable, furnishing their guests with all the conveniences of a metropolitan hotel. Ninety days before, Pithole had been only a spot in the wilderness. Now, it was a city with about 15,000 inhabitants! [26]

The rise of Pithole had been swift and amazing, but its

[24] *Pithole Daily Record*, Sept. 29, 1865.
[25] *Ibid.*, Oct. 14 and Nov. 23, 1865.
[26] Titusville *Morning Herald*, Sept. 20, 1865; *Pithole Daily Record*, Sept. 26, 1865. The population of Pithole has been estimated all the way from 10,000 to 30,000. From comparing contemporary accounts, we may conclude that the population did not exceed 15,000 at its height. Probably the period of greatest gain was the first three weeks of September.

decline was even more breath-taking. When the Homestead well suddenly stopped flowing in August, it caused a sensation among the operators and provided the first shock to the speculative bubble. In November, the old Frazier and the Island well quit spouting. While the failure of these three large wells naturally created a lack of confidence in the territory, many believed that there was little cause for alarm. The completion of the Eureka well on the Holmden farm in November and its production of 1,000 barrels helped to allay the apprehension. Despite its success and that of subsequent wells, an increasing number stopped producing, and more dry holes were drilled. By January, the daily production on Pithole Creek had fallen off to 3,600 barrels; by December, 1866, it was down to 1,800; and by 1867 it was less than 1,000. After having produced 3,500,000 barrels from less than a 100-acre area, the pool seemed to be exhausted.[27]

Simultaneously with the decrease in production, operators, speculators, business men, and others quickly departed from Pithole; and by January, 1866, the eager throng of oily pilgrims had disappeared, and Pithole was a deserted village.[28] A series of disastrous fires razed most of the buildings. The best of those left standing were torn down and moved to neighboring towns, while the others were sold for kindling. The derricks were soon cleared away, the farms reverted to their original state, and today a solemn silence reigns over the grassy land which once provided oildom's greatest excitement.

[27] James Dodd Henry, *History and Romance of the Petroleum Industry*, 286.
[28] Titusville *Morning Herald*, Jan. 22, 1866.

CHAPTER XI

Revolutionizing the Transportation of Oil

THE years 1865 and 1866 witnessed a complete change in the mode of transporting oil to market. Instead of the once familiar piles of barrels at the wells and shipping points and long processions of teams and wagons, the oil region now had miles of pipe line, railroad tracks, and hundreds of railroad tank cars along the sidetracks. Swiftly, but not without opposition, a revolution in the transportation of oil had occurred.

In order to eliminate the risk, waste, expense, and uncertainty of transporting oil by boat or wagon, Heman Janes of Erie proposed at a meeting on Tarr farm in November, 1861, the laying of a four-inch wooden pipe in a trench along Oil Creek and letting the oil gravitate to Oil City.[1] Those present approved the scheme, and James Reed of Erie secured a contract to do the work. Colonel E. C. Clark of Sumner and Clark apparently saw the real possibilities involved and advised applying to the legislature for a general pipe-line charter. Since the Reed contract had not yet been signed, a bill was introduced in the state legislature of Pennsylvania to authorize the building of a pipe line; but on account of the opposition and influence of the teamsters, through their representatives, the first effort to organize a pipe-line company failed. In February, 1862, however, the legislature in-

---

[1] McLaurin, *Sketches in Crude-Oil,* 265.

141

corporated the Oil Creek Transportation Company with a capitalization of $100,000 for the purpose of conveying oil through pipes from Oil Creek to its mouth or to any point on the Philadelphia & Erie Railroad.[2]

The first successfully operated pipe line, used by Barrows and Company of Tarr farm, conveyed oil from one of the wells to their refinery 800 to 1,000 feet away, and it worked admirably over this short distance.[3] During the summer of 1863, Hutchinson and Company laid another from Tarr farm to the Humboldt refinery at Plumer, a distance of about two miles.[4] Through a two-inch pipe, a pump at Tarr farm and two others at intermediate points forced the oil over an elevation of 400 feet above Oil Creek. The line was only partially successful owing to poor pipes, leaky lead joints, which caused a heavy loss of oil, and faulty force-pump machinery. That winter the same parties laid a three-inch cast-iron pipe from the Noble and Delamater well to Shaffer farm on Oil Creek, about three miles away. Again the venture failed for the same reasons. On the other hand, a refiner at Plumer simultaneously laid and successfully operated a two-inch wrought-iron pipe line, three miles long, to the Allegheny River. Early in 1864, a scheme for laying a pipe down the Allegheny to Pittsburgh was proposed, but a large number of the people in the oil region opposed it, for fear it would drive out the teamsters and ruin business; so the project was abandoned.[5] These early experiments demonstrated the practicability of the pipe line, but its successful operation over greater distances did not come until the summer of 1865.

The unsatisfactory condition of the roads, the exorbitant

[2] *Laws of the General Assembly of the State of Pennsylvania, Passed at the Session of 1862, in the Eighty-sixth Year of Independence* (Harrisburg, 1862), 60.

[3] *Oil City Derrick*, Aug. 27, 1909.

[4] Titusville *Morning Herald*, March 6, 1866.

[5] *Derrick's Hand-Book*, I, 35.

charges of the teamsters, and the production of oil faster than it could be hauled away from Pithole influenced Samuel Van Syckel, an oil buyer, to lay a two-inch pipe line from Pithole to Miller's farm on the Oil Creek Railroad, about five miles away.[6] From the moment that Van Syckel got the idea until he completed the project and demonstrated its usefulness, he was the subject of ridicule. Many people believed it to be a visionary scheme and had little confidence in its success. When he talked about it, his friends pitied him; they did all they could to discourage him, told him that it was folly to attempt such a thing, that it couldn't be done, and that it would cost a mint of money. Van Syckel admitted that it might cost $100,000; but he had the money and believed in the idea. Others, not his friends, made him the butt of their ill natured jokes. They would sarcastically inquire, "Do you intend to put a girdle around the world?" "Can you make water run uphill?" Finally he had to take his meals privately at the Morey Hotel as he was unable to endure the scoffings and revilings that greeted him in the public dining hall; and, to avoid the loafers in the front room of the hotel, he would go out and in by the back door.

Van Syckel began laying pipe on September 5. Made to order, the pipe came in sections about fifteen feet long and cost $50 a joint. The joints were lap-welded and tested at a pressure of 900 barrels to the inch. Most of the way it was laid on top of the ground, but for some of the distance a trench two feet deep was dug. While the pipe line was being laid, it was maliciously cut in several places by teamsters who saw a menace to their monopoly and occupation.[7] Along its entire length and in all of the towns of the region placards were posted inveighing against it; and Van Syckel had to station watchmen to arrest anyone tampering with the pipe.

On October 9 the pipe line was completed, and Van Syckel

[6] "Oil Transportation," *Petroleum Age*, VII (1888), 34–36.
[7] Titusville *Morning Herald*, Sept. 27, 1865.

made the first test. Three Reed and Cogswell steam pumps, two at Pithole and one at Little Pithole, forced eighty-one barrels of oil through the pipe in an hour, doing the day's work of 300 teams working ten hours. The experiment worked perfectly. When the fourth pump was added at Cherry Run it was estimated that this quantity could be increased at least 25 per cent. Van Syckel now had the pleasure of seeing his persecutors overwhelmed and silenced.

During the next three months, several other pipe lines were completed to Pithole. A two-inch pipe with a capacity of 2,000 barrels a day began carrying oil on October 24 to Henry's Bend on the Allegheny River, four miles away.[8] Owing to the splendid performance of his first line, Van Syckel laid a second to Miller's farm, and it commenced delivering oil on December 8. The two lines to Miller farm now yielded a revenue of over $2,000 a day.[9] Before the end of December, the Pennsylvania Tubing and Transportation Company had a six-inch pipe in operation between Pithole and the mouth of Pithole Creek, about seven miles away; it was able to deliver 7,000 barrels every twenty-four hours.[10] For the entire distance there was a perfect and constant decline of fifty-two feet to the mile, so that force pumps were not necessary.[11] As a result of the laying of the pipe line, a little town, called Oleopolis, came into existence on the lands of the Baltimore Petroleum Company at the mouth of Pithole Creek.[12]

None of the first pipe lines at Pithole connected directly to the tanks at the wells. Each line had dump tanks to which oil was hauled from the wells and dumped, then pumped away. A. W. Smiley, in charge of buying at Pithole for Van Syckel, hit upon the idea of connecting the lines directly to

[8] *Ibid.,* Oct. 27, 1865.
[9] *Ibid.,* Dec. 11, 1865.
[10] *Ibid.,* Aug. 3, 1865.
[11] *Pithole Daily Record,* Dec. 12, 1865.
[12] Titusville *Morning Herald,* Aug. 3, 1865; *Oil City Register,* Aug. 24, 1865.

the tanks at the wells, and in the summer of 1866, in conjunction with George E. Coutant, built what was known as the accommodation pipe line.[13] It extended over the flats along Pithole Creek, connecting with the tanks at the best wells, and transferred oil from the tanks to the dump stations of the different through lines. Transferring the oil cost twenty-five cents a barrel, whereas teams formerly rendered the same service for fifty cents to a dollar. Within a short time, however, all through lines connected directly with the wells and resulted in the abandonment of the accommodation line, but not before Smiley sold his half-interest for $3,000 cash.

As a result of the introduction of pipe lines, the cost of transporting oil was reduced to the uniform rate of $1 a barrel, which ruined the teamsters' business. No sooner had the first one to Miller farm been completed than teamsters began returning to Titusville and other shipping points with empty barrels; there was no oil to haul. They now faced the stern reality that they were no longer necessary in the oil field. When other pipes were laid to Pithole and the production decreased, over 1,500 teamsters left the city in one week. In order to eliminate the teamsters completely the pipe-line companies kept their prices just below those of the teamsters and yet so high that producers derived little monetary benefit.[14]

With the completion of the pipe lines to Miller farm, Oleopolis, and Henry's Bend, Titusville's importance as an oil shipping center decreased. Once a constant tide of oil teams had rolled into the city, but now fewer and fewer hauled in oil; the receipts had never been so low. Thoroughly aroused over the diversion of the oil trade to Oleopolis and Miller farm, the business men of Titusville belatedly agitated for the building of a pipe line to Pithole. As a consequence,

[13] Smiley, *A Few Scraps, Oily and Otherwise,* 137–139.
[14] Titusville *Morning Herald,* Feb. 5 and 23, 1866.

George J. Sherman, H. E. Pickett, B. Philpot, and others soon organized the Titusville Pipe Company, sold stock, secured a right of way, and in January, 1866, started laying two lines of two-inch pipe in one ditch to Pithole.[15] The costs were: about forty-five cents a linear foot for the two-inch pipe, $1,000 a mile for trenching and laying the pipe, $800 each for force pumps and steam engines combined, and $2,000 for a steam boiler capable of running two pumps. When completed in March, 1866, the pipe line had a capacity of 3,000 barrels a day.

About ten days before the pipe line from Titusville to Pithole had been finished, Henry Harley, formerly a commission dealer in oil, completed two pipe lines from Bennehoff Run to Shaffer farm, a distance of about two miles. Each pipe had a daily capacity of 1,500 to 2,000 barrels.[16] Immediately it drove 400 teams out of the oil region, and those who remained began making threats against the pipe line, forcing Harley to have watchmen patrol the route. One night about two A.M., a fire broke out in Harley's storage tanks at Shaffer, destroying the shipping platform, four tank cars, the tank line, and 450 barrels of oil.[17] Though the cause of the fire was not known, it was generally believed to have been kindled by disgruntled teamsters, so that Harley stationed watchmen around the tanks. Two nights later, about the same hour, a mob of 75 to 100 men, armed with revolvers, rushed down from the woods and underbrush. Before reaching the tanks, they halted. The ringleader notified the watchmen that they intended to destroy the tanks and called upon them to disperse or be shot. When the watchmen retreated, one of the mob threw a fireball into one of the tanks and in an instant the flames spread to the four adjoining ones. The mob remained until the tanks were ablaze, yelling, firing into the air, threatening to shoot any person who interfered; then they returned to the woods and climbed the hill. Fortunately, since the tanks had been almost drained on the previous day, the damage amounted to only about $8,000. The pipes were not destroyed, and within twenty-four hours they were once more discharging oil into the tank cars on the railroad. Unsuccessful in destroying the pipe line, the teamsters now sent an anonymous letter to the superintendent of the Bennehoff Petroleum Company, stating that unless the company discontinued supplying oil to the Harley line, the wells would be burnt.

Van Syckel had similar trouble. The teamsters either broke the pipe line apart with pickaxes or fastened log chains around the pipes, hitched on with horses and pulled them apart. To protect the line, he sent to New York for carbines and provided an armed patrol. Rather than resort to further violence, the teamsters now reduced the price of teaming; but their ruin was inevitable, for the pipe line was more efficient and it could always carry the oil cheaper than the teamster in order to eliminate him.

Not long after the completion of Van Syckel's pipe line, his two partners failed financially, forcing the First National Bank of Titusville to press its claim against them. Since they owed the bank $15,000, Van Syckel assumed payment of the debt and agreed that the bank should take over the line and operate it until the debt was liquidated, which it did.[18] On account of unforeseen financial difficulties, Van Syckel never regained control, and the line was sold to W. H. Abbott and Henry Harley. In the fall of 1867 they purchased the charter of the Western Transportation Company, which had laid a pipe line from the Noble and Delamater well to Shaffer; under its charter, Abbott and Harley combined their lines to Pithole and Bennehoff Run to form the Allegheny Transportation Company, the first great pipe-line company.[19]

Another notable improvement in the facilities for trans-

18 "Oil Transportation," *Petroleum Age*, VII (1888), 34–36.
19 J. T. Henry, *The Early and Later History of Petroleum, etc.*, 527–528.

porting oil was the appearance of the Empire Transportation
Company in the oil region in the spring of 1865. It was
organized by the Pennsylvania Railroad for the purpose of
securing a larger share of the oil traffic originating along
Oil Creek for the Philadelphia & Erie, which it then operated
under lease. About ten railroads connected with the Philadelphia & Erie to form direct routes to the east and west,
but each one had its own time tables, rates, and methods of
doing business.[20] A shipper using this route had to make arrangements with each road, and frequently his freight had to
be changed at terminals from one car to another owing to
the differences in gauge. Because of these inconveniences,
the New York & Erie Railroad, controlled by "Jim" Fisk and
Jay Gould, and the Atlantic & Great Western, a part of the
Erie system, carried most of the oil to market. Acting as an
intermediary between the shippers and the various railroads
connecting with the Philadelphia & Erie, the Empire opened
its office at Shaffer farm in June, 1865, and offered cheap,
fast, and reliable service. It furnished its own cars and
terminal facilities, quoted an all-rail rate to the east, and
collected money due shippers. For these services, it received a commission on the business secured for the Philadelphia & Erie, a rental fee for the cars and the other facilities
it provided.

At first shippers hesitated to patronize the Empire line,
for they did not want to incur the ill will of the Oil Creek
Railroad.[21] However, the demand for cars to haul oil was so
great with the opening up of Pithole, and the Empire could
furnish them so regularly and in such large numbers that
shippers soon began to use its services; and the company developed a flourishing business, which incurred the hostility
of the Erie management. Afraid that Fisk and Gould were

[20] Ida M. Tarbell, *The History of the Standard Oil Company* (New York,
1925) , I, 24.
[21] *Derrick's Hand-Book*, I, 963.

attempting to secure possession of the pipe lines in order to cut off the supply of oil to the Empire, the latter purchased the Titusville Pipe Company in June, 1866, as a protective measure and went into the pipe-line business. In time, it acquired other pipe lines; and by the end of the decade the Empire had become one of the most efficient business organizations in the oil country.

Simultaneously with the laying of the pipe lines, there was a notable extension of the railroads into the oil region. From the north, the Philadelphia & Erie Railroad began early in 1866 to build a branch line from Irvineton to Pithole while the Warren & Franklin Railroad commenced laying tracks from Warren to Oleopolis.[22] Most of the new construction, however, began in or around Oil City and extended northward into the oil field. The Kersey Mineral and Oil Company started building a railroad from the mouth of Cherry Run to Plumer. Another road was built from Oil City to Oleopolis to connect with the Oleopolis & Pithole Railroad, which had been completed in December, 1865. Thus a direct connection was established in the spring of 1866 between Pithole and the Atlantic & Great Western at Oil City, and through shipments could now be made to New York or elsewhere. When the Warren & Franklin Railroad finished its road to Oleopolis in July, through shipments could be made to the east in the opposite direction over the Philadelphia & Erie. The Warren & Franklin then purchased the Oil City and Oleopolis line, giving it access to Oil City. The Farmers' Railroad began laying its tracks from Oil City to Petroleum Centre early in 1866, but the Warren & Franklin purchased a majority of the stock in August, consolidated it with the rest of its holdings, and by October had the road completed into Petroleum Centre.

[22] Larson, *Jay Cooke*, 153. Jay Cooke, J. Edgar Thomson, Thomas A. Scott, and William G. Moorhead were associated in building the Warren & Franklin Railroad.

The most remarkable of the early railroads in the oil country was the Reno & Pithole Railroad. Commencing at Reno, where it connected with the Atlantic & Great Western Railroad, the company started early in 1866 to build its line to Pithole. Under the direction of General A. E. Burnside, an army of over 1,500 laborers graded and laid the rails. From the appearance of the grade over the hills, one could not determine whether it was intended for a turnpike or a railroad; the grade varied from eighty to one hundred feet a mile and was uphill all the way to Rouseville.[23] It was so steep that the first locomotive was unable to ascend the grade at Reno, and a change had to be made. In several places, trestles forty to seventy feet high had to be built across ravines, the one across Oil Creek being about sixty feet high. By February, 1866, the track was completed to Rouseville and trains began running.

Owing to the ruinous rates, the inadequate service, and the monopolistic character of the Oil Creek Railroad, a group of Titusville citizens organized the Oil Creek Lake and Titusville Mining and Transportation Company for the purpose of building a competing railroad to Union Mills.[24] To pay for the railroad, the stockholders estimated that they would need only one-third of the business of the Oil Creek Railroad for a year. With a small army of laborers, the company started work late in February, 1866; but it did not finish until 1871.

As a result of the railroad building during 1866, numerous short, independent railway lines had been constructed throughout the region, which aided materially in improving the facilities for transporting oil. In January, 1868, representatives of the Warren & Franklin, the Farmers', and the Oil Creek Railroad met in Philadelphia and merged the three railroads into the Oil Creek & Allegheny River Railroad with

23 Oil City Register, Jan. 11, 1866.
24 Titusville Morning Herald, Feb. 8 and 21, 1866.

a capitalization of $4,250,000.[25] When this consolidation had been completed, the new company purchased the Reno & Pithole Railroad. Thus all the small railway lines in the oil region had been combined into one.

The railway construction during 1866 made possible the extensive introduction of the oil tank car, which Amos Densmore, a member of the firm of Densmore Brothers, buyers at Miller farm, devised during the summer of 1865.[26] On an ordinary flatcar, he built two wooden tanks, one on each end over the trucks, filled each with forty-two to forty-five barrels of oil, and started the car on its way to New York. From Corry, Salamanca, Hornellsville, Elmira, and other points along the way, messages were sent back stating that the tanks were not leaking. The experiment having proved a success, the Densmores constructed other tank cars and soon had quite a number in service.

Early in 1866, influenced by the experimentation of the Densmores, the Empire Transportation Company sent to its agent at Shaffer a boxcar within which three wooden tanks had been erected.[27] It instructed the agent to be very careful, for the whole thing was an experiment and it was expected that the car would be restored to other use. When partially filled with oil, the tanks started leaking, and in order to tighten the hoops workmen had to tear off the side and end boards of the car; it was virtually ruined, except for use in hauling oil. After much hard work and the loss of considerable oil, the leaks were stopped, the tanks filled, and the car forwarded to its destination. Within a short time, however, the Empire line adopted the type of tank car as used by the Densmores, and in the spring of 1866 it had several hundred in use.

Other shippers quickly resorted to the same method. It

[25] *Ibid.*, Feb. 6, 1868.
[26] Smiley, *A Few Scraps, Oily and Otherwise*, 145–146.
[27] *Derrick's Hand-Book*, I, 964.

became customary for individuals to furnish the tanks and the railroads the flatcars on the condition that they should be constantly employed.[28] Usually they placed on each car two forty-barrel tanks, built of clear pine plank and provided with closely fitted covers. They were firmly attached side by side to the floor. These tank cars became a permanent feature in 1866, and hundreds of them came into use. This type was used until the iron boiler tank car was adopted by the Empire line and first put into use in February, 1869.

Although the introduction of the tank car constituted a distinct improvement in the method of shipping oil, it brought the whole oil trade under the control of the railroads, which established a monopoly of the worst sort, dictated prices to producers, and drove shippers from the field.[29] Now the producers had to fight not only the pipe lines, but the railroads and tank-car companies.

[28] Titusville *Morning Herald,* March 9, 1867.
[29] *Ibid.,* July 14 and Dec. 15, 1866.

ABBOTT & HARLEY'S PIPE-LINE TERMINAL, SHAFFER FARM, 1866

EARLY TANK CARS

CHAPTER XII

The Depression Years

EVEN before Pithole reached its zenith, a number of influences, domestic and foreign, conspired to produce a general business depression throughout the oil region which caused a drastic curtailment of operations in 1866 and especially in 1867. Abroad, the expiration of Young's patent in England and Scotland for manufacturing kerosene from Scottish shales had reduced the price and made it difficult for petroleum from the United States to compete. Although not generally known in the oil field, the amount of oil sent from New York to Cork, Ireland, in 1864 and 1865 constituted an important part of our export trade, and the outbreak of the Fenian rebellion paralyzed the trade and caused a heavy decline in our exports.[1] Then, with the rapid development of the petroleum industry in Canada, Europe, the Far East, and South America, where conditions favored the cheaper production of oil, the United States government issued a report relative to the progress in each country, and that had a depressing effect.[2] At home, the restoration of the national currency at the end of the Civil War to something like its real value affected every business in the country, and the petroleum industry did not escape. Added to this was the fact that eastern speculators always endeavored to depress the price at the wells in order to buy oil for little or nothing.[3] As a result of all these factors operating collectively, the price

[1] Titusville *Morning Herald*, March 13, 1866.
[2] *Ibid.*, March 17, 1866.
[3] *Ibid.*, March 28, 1866.

of oil dropped from $7.50 in March, 1865, to $2.50 a year later. Though the price of oil had been considerably reduced, the freight charges continued with little or no change. The cost of shipping oil from Shaffer to Corry amounted to 85 cents a barrel, and from Corry to New York, to $2.04, or a total charge equal to almost one-fourth of the New York sale price.[4] The Atlantic & Great Western Railroad reduced its tariff about half, but even with the reduction it did not afford a sufficient margin to ship oil to New York when the price was so low. Finally, the federal tax of $1 a barrel on all crude oil also weighed heavily upon the industry. Passed by Congress in March, 1865, when oil was $7.50 a barrel, the tax did not adversely affect the producer, but with the decline in price, it proved burdensome, especially to the owners of pumping wells yielding ten to twenty barrels.

In view of the general situation, hundreds of small wells were rapidly abandoned early in 1866, and the daily production fell off by half; according to one estimate, drilling stopped on nearly 8,000 wells. The whole region was covered with abandoned derricks; very few operators commenced any new wells; and scores of small refineries began shutting down.[5] A stagnation of business set in throughout the oil field, the natural consequence of which was to throw thousands of men out of work.

Locally, the severity of the depression was intensified by the announcement of the failure of the banking house of Culver, Penn and Company of New York on March 27, 1866, on account of its heavy speculative activity in building the town of Reno and the Reno & Pithole Railroad.[6]

---

[4] *Ibid.*, March 22, 1866. The total cost of delivering a barrel of oil in New York amounted to $9.14, or twenty-two and three-fourths cents a gallon. It sold in New York for twenty-four cents a gallon.

[5] *Ibid.*, April 23, 1866.

[6] *Ibid.*, March 28 and 29, 1866. The Reno Oil and Land Company was organized in Dec. 1865. It purchased 1,200 acres of land on the Allegheny River, four miles from Oil City, and proceeded to lay out a model town, Reno, and drill wells.

Headed by C. V. Culver, member of Congress from Pennsylvania, Culver, Penn and Company owned an interest in a string of banks in the oil region at Titusville, Franklin, Meadville, Oil City, and Petroleum Centre; and its failure resulted in closing five banks and in a heavy loss to depositors.

With the control of the market beyond the producers' reach and with no change in freight rates possible until competing lines were built, one source of relief seemed to be a repeal of the tax on crude oil. The United States Revenue Commission had already advocated the repeal of the tax, but its report was only a recommendation to Congress.[7] Vigorous action was necessary, and so the producers now united in the greatest cooperative effort ever seen in the oil region. After protest meetings at Petroleum Centre and Pithole, the Titusville Board of Trade invited the producers from all over the region to meet in that city on February 1 to discuss all their grievances.[8] Among other things, they adopted a resolution in favor of repealing the tax and authorized J. T. Sawyer of Pithole and E. W. Mathews of Titusville to go to Washington and present their views to the Ways and Means Committee of the House of Representatives.

In Washington, Sawyer and Mathews found that those Congressmen upon whom they called lacked any information about the oil business, which was probably due to the fact that C. V. Culver, who represented the oil district in Congress, had been absent much of the time from his duties.[9] They talked with Justin Morrill, chairman, and the other members of the Ways and Means Committee, and each one gave assurances of his willingness to bring about a repeal of the tax. In James A. Garfield, who owned some oil lands, they found a champion for their cause.

Upon their return from Washington, Mathews and Sawyer

---

[7] Hayes, "Report of the United States Revenue Commission on Petroleum as a Source of National Revenue," *House Exec. Doc.* No. 51, 39 Cong., 1 sess., 39.

[8] Titusville *Morning Herald,* February 1 and 2, 1866.

[9] *Ibid.,* April 9, 1866.

gave a report on the situation before another public meeting at Titusville on April 20.[10] Once again the producers passed resolutions demanding a repeal of the tax, authorized George M. Mowbray and W. H. Comstock to present the memorial to Congress, and voted that Culver should resign or enter upon his congressional duties. The concerted efforts of the oilmen brought the desired result, for under the leadership of Garfield, Morrill, and Senator Edgar Cowan of Pennsylvania, Congress repealed the tax on May 9, making the law effective at once.[11]

Though the repeal of the tax led owners of small wells to begin pumping once more and drillers to resume work, it did not cure the trouble, and additional influences continued to keep the price of oil low.[12] A large amount of oil had been exported earlier in the year, flooding European markets, and now the outbreak of the Austro-Prussian War removed the hope of any large export demand. The general introduction of the torpedo enlarged the supply, without a corresponding increase in consumption, and adversely affected the market.[13] Finally, one of the most persistent causes of the continued low price of oil was the constant flooding of eastern markets by the new tank-car operators.

To combat the situation, some of the producers wanted to organize a combination for the purpose of erecting storage tanks along Oil Creek and regulating the price of oil; but most of the producers had come to make money quickly and then depart, and so were not interested in the solution of a common problem. Individuals and firms, however, pro-

[10] *Ibid.,* April 17 and 21, 1866.
[11] *Congressional Globe,* 39 Cong., 1 sess., part III, 2433, 2446, 2471, 2482.
[12] Titusville *Morning Herald,* July 6, 12, and 14, 1866.
[13] The long-talked-of experiment of exploding a torpedo in a well to enlarge the production was tried in the Ladies well on Watson's flats, below Titusville in Jan., 1865. Eight pounds of powder were lowered into the well and touched off at a depth of 463 feet. Within a relatively short time, this procedure was universally used to increase the production of new as well as old wells.

Mather Photograph

WASHINGTON STREET, PETROLEUM CENTRE, 1868

PIONEER, OIL CREEK, 1865

ceeded to erect iron storage tanks; and by the end of the year tankage for 280,000 barrels had been constructed along Oil Creek and at Oil City, 50,000 barrels at Oleopolis, and over 100,000 at Tidioute.[14] Other producers began closing down their wells to wait for a better price. By the end of 1866, the daily production had been materially reduced and relatively few new wells were being drilled.

In spite of the depression during 1866, new territory was developed on West Hickory Creek and Dennis Run, tributaries of the Allegheny. With oil found at a depth of 200 to 300 feet and less expensive to secure, West Hickory had an immediate advantage over Oil Creek or Pithole. Though the wells were small, the region produced a rich, heavy lubricating oil, more valuable than ordinary oil. On Dennis Run there were some of the finest pumping wells in the region with a greater yield than those on West Hickory. About the middle of July, the daily production amounted to 2,725 barrels.[15] In the confusion and excitement of the summer, a new town called Babylon rose and, like its ancient prototype, became one of the "toughest" places in the whole oil region with fights, brawls, murders, and vice of all kinds.

While West Hickory and Dennis Run attracted considerable attention, the region around Petroleum Centre on Oil Creek furnished the greatest excitement and became the largest producing area in 1866. Some New Yorkers had purchased the farm of George Washington McClintock on Oil Creek in 1859 for a few hundred dollars, organized the Central Petroleum Company of New York, and drilled the first well in 1860.[16] As other wells were drilled, a small village developed along the creek, and since it was about halfway between Titusville and Oil City, it came to be called Petroleum Centre.

---

[14] Titusville *Morning Herald*, Nov. 19, 1866.
[15] *Ibid.*, July 19, 1866.
[16] *Ibid.*, Dec. 1, 1866.

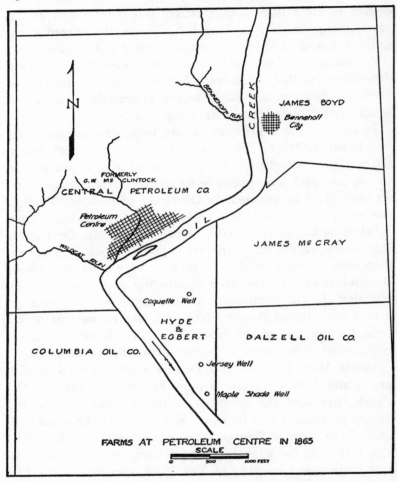

FARMS AT PETROLEUM CENTRE IN 1865
SCALE

North of the village about a quarter of a mile was the mouth of Bennehoff Run; the land at the mouth and for a quarter of a mile up the run was a part of the McClintock farm. The adjoining farm belonged to John Stevenson. In the spring of 1865 the Phillips brothers purchased Stevenson's farm for $35,000.[17] They formed the Ocean Oil Com-

[17] *Ibid.*, Oct. 23, 1866. They were C. M. and T. W. Phillips.

pany and commenced drilling a well on an elevation about 250 feet above Oil Creek. Late in August, the Ocean well, as it was called, began flowing 300 barrels. This well exploded the theory that producing wells could not be drilled on high hills, and the immediate effect of the strike was to attract operators to Petroleum Centre. Half-acre leases in the surrounding farms sold for $500 to $3,000. A sixteenth interest in new wells being drilled brought $50 to $2,000. Test wells were watched with great interest, and the excitement rose to fever heat. By December, over 150 wells were being drilled.

In the succeeding months, the Stevenson farm, the land on Bennehoff Run, and that belonging to the Central Petroleum Company vied with one another to see which would lead the excitement. No sooner did one produce a large well than another would develop a larger one. Early in 1866 the daily production on the Stevenson farm jumped to 1,600 barrels, while that on Bennehoff Run amounted to 1,200.[18] Late in February the production of Bennehoff Run greatly increased on account of the completion of the celebrated Phil Sheridan well; it flowed 1,000 barrels and was the largest in the region. A dividend of $1,000 a month for each sixteenth interest was declared on this well for the next four months.[19] A short time later, the Philadelphia well commenced flowing 800 barrels. In general, the producing wells at Petroleum Centre were neither so famous nor so large; but they were numerous, and those on high elevations yielded a greater quantity than those on the flats. The mountainous territory seemed pregnant with oil.

Surrounded by some of the best oil-producing farms, Petroleum Centre was suddenly transformed within a few weeks from a village of 500 into a lively town with 3,000 population, a bank, two churches, a theater, a half-dozen hotels,

[18] *Ibid.*, Jan. 10, 1866.
[19] *Ibid.*, Aug. 11, 1866,

a dozen dry-goods stores, three or four livery stables, scores of offices for brokers, shippers, and producers, dance halls, saloons, gambling dives, and boarding houses.[20] With the influx of all sorts of people and an absence of local government, the town soon acquired the reputation of being very wicked, even worse than Sodom and Gomorrah, according to some people. Of the character of its people, one visitor said: "It was composed of all classes, from the murderer to the minister of the gospel. The thugs, gamblers and soiled doves were in the majority. About 200 of the latter came down from the fast-waning Pithole City and took up their abode in Petroleum Centre's dance houses, of which there were a half dozen, free and easies and other 'houses.' The male population was but little better than the female and, for this reason, Petroleum Centre was noted as a 'daisy town.' "[21] For pure unadulterated wickedness, it eclipsed any other town. Thugs so terrorized the place that it was not safe to walk the streets at night; houses of prostitution flourished; the innocent were fleeced, sandbagging seemed to be a regular occupation, and gambling houses ran wide open.[22]

The extension of the Oil Creek Railroad to the Boyd farm, opposite Petroleum Centre, in July, 1866, stimulated the growth of another oil town, and the place became a beehive of activity. In an incredibly short time Boyd's farm was transformed from a miserable wilderness into the most important point on the creek.[23] Because it was near the mouth of Bennehoff Run, many people called the town Bennehoff City.

During the summer the wells on Pioneer Run, a small stream running through a deep gorge and emptying into Oil Creek a short distance above Bennehoff Run, produced the largest amount of oil. Pioneer Run was almost unknown as

[20] *Ibid.*, Feb. 21, 1866. Counting those on the surrounding farms, the population amounted to about 6,000 (*ibid.*, March 8, 1866).

[21] Brown, *Old Times in Oildom*, 16.

[22] Steele, *Coal Oil Johnny*, 169.

[23] Titusville *Morning Herald*, July 20 and Aug. 6, 1866.

DWELLING HOUSE AND INHABITANTS, PIONEER RUN

an oil-producing territory until the completion of the Lady Brooks well early in April, 1866, which flowed over 600 barrels a day. Speculators now eagerly sought the side hills, operators began putting down other wells not only on Pioneer Run but on Western Run, a tributary stream; and by summer the daily production amounted to 2,500 barrels.[24] As in similar rushes, a small mushroom town, Pioneer City, rapidly developed at the mouth of Pioneer Run.

By fall, the production of Bennehoff Run jumped to 2,200 barrels, which surpassed the output of any other district. Some of the largest producing wells were on John Bennehoff's farm, which lay across Bennehoff Run, Western Run, and Pioneer Run. For many years Bennehoff, an illiterate, poverty-stricken old German farmer, had wrung a bare subsistence from the stony soil. As most of his farm was highland territory, operators did not look upon it with favor until the fall of 1865, when one completed a well which produced 300 barrels. Bennehoff was now besieged by others who wanted leases. Small lots were taken at boom prices and one-fourth the oil, and shortly a dozen wells were being drilled all over his place.

Endowed with unusual business shrewdness, the son, Joseph, persuaded his father to give him the management of leasing. Joseph increased the royalty to one-half and found eager lessees as large producing wells were completed. Riches were thrust upon John Bennehoff, and soon he had an income estimated at $6,000 a day; his royalty in November, 1866, at a time when the price of oil was low, exceeded $33,000.[25] Distrustful of bankers and banks, the father and son purchased two safes in which to keep their wealth at home. The whole community knew that they kept their money in the house, and friends advised using the bank, but the Bennehoffs did not listen. They employed a night watch-

[24] *Ibid.*, July 30, 1866.
[25] *Ibid.*, Jan. 5, 1867; *Derrick's Hand-Book*, I, 88.

man—with instructions not to admit anyone other than the family after nightfall.

Late in 1867, Louis Waelde, Jacob Shoppart, George Miller, and James Saeger, all residents of Saegerstown, Pennsylvania, laid plans to rob the Bennehoffs. To assist them, Saeger, the chief conspirator, engaged the services of two professional burglars from Philadelphia. With everything in readiness, Saeger and the Philadelphians proceeded to execute the plan on the evening of January 16, 1868. About seven o'clock, while John Bennehoff, his wife, his niece, and the watchman, George Geiger, were sitting around the kitchen stove, some one knocked on the door.[26] Over Bennehoff's protest, Geiger opened the door and three masked and armed men rushed in, threatening to shoot anyone who made a noise. Though over sixty-five years old, Bennehoff attempted to resist; the robbers knocked him down, bound him, rifled his pockets, secured the key to one of the safes, and then tied Mrs. Bennehoff, the niece, and Geiger. With the key, they opened one safe and took about $210,000—about $50,000 in government bonds and the rest in greenbacks.[27] Unable to open the other, the robbers waited for the return of Joseph Bennehoff from church, for he could work the combination. When he did not come within an hour, they forced Geiger to harness and hitch a horse to a cutter, then rebound him, and hurriedly departed. Bennehoff's niece soon got loose and released the others, and the hired man rushed to call the neighbors. The news of the sensational robbery quickly spread and became the theme of conversation throughout the oil region for weeks.

The next day Joseph Bennehoff offered a reward of $10,000, which was ultimately boosted to $50,000, for the apprehension of the criminals, and the family spared no expense to

26 Titusville *Morning Herald*, Jan. 17 and 18, 1868; James Dodd Henry, *History and Romance of the Petroleum Industry*, 301–305.

27 Titusville *Morning Herald*, Jan. 18, 24, and 25, 1868. Reports concerning the amount stolen ranged as high as $500,000.

catch the culprits, spending no less than $15,000 during the next six months.[28] In July, 1868, Geiger, Waelde, Shoppart, and Miller were apprehended and tried for the robbery at Franklin.[29] Geiger was acquitted owing to a lack of evidence, but the others were found guilty and served prison terms.

The Bennehoffs never recovered the stolen money and never secured the conviction of James Saeger, who planned and successfully engineered the whole affair. As soon as the robbery was over Saeger gave $1,300 to each of his accomplices, Miller, Shoppart, and Waelde, $25,000 each to the two professionals from Philadelphia, and kept the rest. The two Philadelphians escaped to Canada. Saeger fled to Havana, Cuba, then to the cattle fields of Texas.[30] In time, he acquired a large herd of cattle and became widely known, under an assumed name, as a successful and substantial cattle dealer. Though finally apprehended in Denver in 1874, he was not returned for trial on account of the expense involved and a fear that he could not be transported safely across the plains.

As the severity of the depression increased during 1867, even West Hickory, Dennis Run, and the region around Petroleum Centre felt its full effects. At Petroleum Centre, where there were approximately 2,000 wells, many were abandoned; about one well in six was in operation.[31] From Oil City to Tidioute, there were innumerable derricks and engines, but only one well in fifty produced any oil.[32] Generally speaking, all work had been stopped; and the region looked desolate. Hundreds of houses were tenantless, and villages of fifty to one hundred houses were silent. Advertisements of sheriff's sales filled many pages of the *Venango*

[28] Titusville *Morning Herald*, Aug. 8, 1868.
[29] *Ibid.*, July 29 and Aug. 8, 1868, and Feb. 1, 1869.
[30] James Dodd Henry, *op. cit.*, 307; *Titusville Morning Herald*, April 30 and May 14, 1874.
[31] Titusville *Morning Herald*, Oct. 18, 1867.
[32] *Ibid.*, Aug. 14, 1867.

Spectator and told the end of the story for scores of companies. Discouraged by losses and the dark prospects for the future, many left the oil region never to return. Others erected more iron tankage in which to store oil for higher prices; and by the middle of the summer there were sufficient tanks to hold more than 800,000 barrels.[33]

Beginning in 1868, the price of oil steadily advanced from $1.55 in January to $4.72 in July; the cost of drilling and operating wells was reduced by decreasing prices for machinery, and new fields continued to be opened. A small well near Bradford, Pennsylvania, was completed, but no one attached any importance to it; operators were attracted in greater numbers to Butler and Clarion counties; but the opening up of new fields at Shamburg and Pleasantville provided the greatest excitement and helped to throw off the lethargic condition which had prevailed in the oil region for over a year.

Several wells had been started on upper Cherry Run early in 1865, but a lack of confidence in the territory and the rise of Pithole handicapped its development. Late in 1865, however, Dr. G. S. Shamburg of Pittsburgh, superintendent of the Pittsburgh and Cherry Run Petroleum Company, commenced drilling a well on the Stowell farm, about two miles southeast of Miller Farm, three miles south of Pleasantville, seven miles from Titusville, and four miles northwest of Pithole.[34] Completed in February, 1866, it pumped 25 barrels at first, then 125. Although the Shamburg well opened up a new field, the depression prevented any extensive operations from being undertaken. During the summer of 1867, however, the drilling of the Cheney well on the Atkinson farm proved that oil was not confined to the immediate vicinity of the Shamburg well. Several other wells were drilled, and a little settlement soon came into existence with

[33] *Ibid.,* July 9, 1867.
[34] *Ibid.,* Oct. 18, 1867.

BENNEHOFF RUN, 1865

THE SHAMBURG FIELD, 1868

a population of 200. Although it was commonly called Shamburg, the post office was designated as Champion.

The completion of the Jack Brown well on the Atkinson farm, which flowed 500 barrels, started a rush to Shamburg in December, 1867.[35] On account of this well and the Fee, which commenced flowing in February, another village, called Atkinson, began to develop around the Cheney well. Instead of erecting new houses or business buildings, people simply moved many of them from Pithole. Between Shamburg and Atkinson, another town called Backus City, sprang into existence on the Tallman farm. While hundreds of buildings were under construction all over the Shamburg area and new wells were being drilled, the excitement and speculation was considerably less than they had been at Pithole. On the other hand, no point in the oil region attracted so much attention at the time as upper Cherry Run.

Reaching a daily production of about 2,500 barrels in June, the Shamburg field gradually declined, and by November had fallen below a thousand. The producing belt included only a few acres. More important as a factor in the decline of the Shamburg field was the excitement which had developed at Pleasantville, for it was the natural thing to follow the crowd from one big strike to another.

One day, late in the fall of 1867, Abram James, an ardent spiritualist, was driving from Pithole to Titusville with three friends.[36] A mile south of Pleasantville the "spirit-guide" caused him to jump out of the conveyance and leap over the fence into a field on the William Porter farm. Hurrying to the north end of the field, James fell to the ground, marked the spot with his finger, thrust a penny into the dirt and fell back pale and rigid. Restored to consciousness, James told his friends of a revelation that streams of oil lay beneath the soil.

[35] *Ibid.*, May 16 and Nov. 7, 1868.
[36] McLaurin, *Sketches in Crude-Oil*, 164–165; Titusville *Morning Herald*, Feb. 3, 1868, and May 26, 1869.

He leased the property, borrowed money, raised a derrick over the spot where the penny lay, and commenced drilling amid the scoffs of unbelievers. When down 700 feet and past the third sand rock, he became the laughingstock of the region; but "Crazy James" kept on drilling. When he proceeded to build tanks to receive the expected oil, people laughed louder. Early in February, 1868, James struck oil, and his well, called the Harmonial, in honor of the spiritual philosophy, pumped 130 barrels.

Though a small well, the Harmonial opened up the Pleasantville field and started what promised to be a second Pithole. Operators at once rushed in to secure leases on adjoining farms, and the fact that experienced oilmen were willing to pay $500 to $1,250 an acre created confidence in the territory. Shortly, at least fifty wells were being drilled; and at the end of June the daily production amounted to more than 600 barrels.

Commencing in July and continuing for the next month or six weeks, hundreds of men flocked to Pleasantville. Of the rush, one visitor wrote: "Pleasantville is booming! or, in other poetical phrase, she's red hot! The crowds yesterday were larger than ever. Thirty-two stage loads of passengers went from Titusville between the hours of seven A.M. and three P.M." [37] All day long a continuous stream of vehicles moved from Titusville to Pleasantville. Moreover, numerous express wagons moved slowly along the road, loaded with merchandise and household goods.

From daylight until dark the streets were thronged with men eager to buy or lease oil land, and the completion of new and successful wells stimulated the excitement. Though it produced only 100 barrels, the Fisher Number 2 on the Hibbard farm on August 2 created a sensation, for it opened up 100 acres of untested land at the eastern edge of the borough and resulted in a greater demand for leases. The

[37] Titusville *Morning Herald,* Aug. 12, 1868.

most productive well was the Grant Number 1, only a few rods from the Harmonial, and it yielded 150 to 200 barrels. On the opposite side of the road was the Bean well, remarkable for its tremendous flow of gas, which escaped from a tube extending twenty feet in the air. When lighted at night, it illuminated all the surrounding fields. Struck on August 7, the Goss and Carll well on the Herbert tract constituted the champion well, producing 220 barrels. By the end of August, about fifty wells had been completed, and the average daily production at Pleasantville increased to over 2,000 barrels.[38] At the same time, the producing belt had been pretty well outlined—an area about two and a half miles wide and three to four miles long.

In two months' time, the population of Pleasantville had jumped from 1,000 to 3,000, and this quiet and peaceful little village had been transformed into a bustling oil town.[39] Great rows of shanties and large frame tenements began to appear among the neat little white houses, making a curious mixture of the old and new. Pleasantville possessed many other characteristics of pioneer oil towns, but none of the repulsive features. An established civilization preceded the oil rush, and the high moral standards of the residents kept the community relatively free from vice of all sorts. Instead of a saloon at every third door, there was only one place where liquor could be purchased; rowdyism and drunkenness did not exist; and disreputable amusements and houses of prostitution were not tolerated.

The main producing area at Pleasantville soon became so thickly covered with pumping wells that it no longer offered any inducement for investment. Moreover, drilling wells so close together, where the oil-bearing rock was but thirty feet thick, resulted in a steady decrease in production. As the wells became exhausted, the excitement rapidly de-

[38] *Ibid.*, Aug. 22, 1868.
[39] *Ibid.*, July 22, 1868.

creased in September, the crowd began deserting Pleasantville; and by the end of 1868 the boom had collapsed. For a
territory that had created so much excitement, had made so
much noise, and upon which so much money had been expended in its actual development, probably none produced so
little oil as the Pleasantville field.

Relieved of the turmoil and confusion, the residents and
property holders began removing the abandoned derricks,
restoring sidewalks and streets, and erasing the grease spots.
Within a relatively short time, the village reverted to its previous state with elegant homes, clean and well graded streets,
and fine shade trees.

SPRING STREET, TITUSVILLE, 1865, SHOWING
THE OLD AMERICAN HOTEL

J. A. MATHER AND HIS BOAT ON OIL CREEK

CHAPTER XIII

Oil Towns That Lived

BETWEEN 1859 and 1869 the oil region witnessed the rise and decline of many oil towns like Pithole, Petroleum Centre, Shamburg, Babylon, and Pioneer. They sprang into existence on account of the completion of some important well, and they disappeared overnight without leaving anything worth while by which to be remembered, except that each had been a "red-hot" town for a few weeks or months. Of a much different character was the history of Oil City and Titusville. The growth of each was as phenomenal as that of other oil towns, but they survived and emerged at the end of the decade as the largest towns of the oil region.

For years prior to Drake's discovery, the site of Oil City had been a favorite overnight stopping place for hundreds of raftsmen, who ran lumber rafts to Pittsburgh from Coudersport, Warren, Tionesta, and other points up the Allegheny. By 1859 a little village had developed, which was called Cornplanter after the renowned chief of the Seneca Indians; it consisted of a store, a gristmill, several dwellings, and the Moran House.[1]

With the striking of oil, Graff, Hasson and Company, which owned about 1,000 acres of land on the east side of Oil Creek with an extensive front along the Allegheny, laid out lots for a town. On account of the rush the village rapidly grew, and in 1861 the name of the post office was changed from Cornplanter to Oil City. In January, 1862, the state

[1] Brown, *Old Times in Oildom*, 34–35; *Oil City Register*, Nov. 17, 1865.

legislature made it a borough, and the first newspaper, the *Oil City Register*, was established. In the spring of 1863 Charles Haines and Joseph H. Marston purchased that portion of the east side of Oil Creek now known as Cottage Hill, laid out lots, and the buyers soon erected some neat dwellings. A mania for building prevailed, and the sounds of the hammer and saw could be heard everywhere from morning until night. No fewer than 100 buildings were under construction in June, 1864, and the contracts had been let for many more.[2] As in other oil towns, the buildings were hurriedly constructed without any thought of architectural beauty or permanence.

Because of its advantageous location at the junction of Oil Creek and the Allegheny, many early oil buyers made their headquarters at Oil City, while their agents scoured the country for oil. In the fall of 1863 the Michigan Rock Oil Company, which owned the land on the west side of Oil Creek at Oil City, divided its extensive river front into lots suitable for oil landings and storage yards. Different firms bought them, and soon warehouses, oil offices, supply stores, tool shops, and cooper shops crowded the river front from the mouth of Oil Creek to the Moran House. By October there were at least twenty oil landings, and the capital invested amounted to about $10,000,000.[3]

As Oil City developed, smaller towns came into existence around it. William L. Lay and Company bought the site opposite the mouth of Oil Creek in 1863 and founded the town of Laytonia. The next spring James Bleakly of Franklin and others purchased the Downing farm, adjoining the Lay property, not so much for drilling oil as for speculation in city lots. Shortly the town of Imperial began to rise. On the river about a half-mile above Oil City, another village, known as Siverlyville, came into existence.

[2] *Oil City Register*, June 9, 1864.
[3] *Derrick's Hand-Book*, I, 34.

By the spring of 1865, Oil City, including Laytonia and Siverlyville, had a population of 10,000! [4] There were nearly a hundred stores and workshops of various sorts, fourteen hotels, a public school, four churches, four banks, eight doctors, and about fifty liquor dealers. Of the city's appearance, one traveler wrote: "Oil City at last. Oil City with its one long, crooked and bottomless street. Oil City with its dirty houses, greasy plank walks, and fathomless mud. Oil City, where horsemen ford the street in from four to five feet of liquid filth, and whose inhabitants wear knee boots as part of indoor equipment. Oil City, which will give the filthiest place in the world three feet advantage and then beat it in depth of mud. . . . Oil City is worthy of its name. The air reeks with oil. The mud is oily. The rocks hugged by the narrow street, perspire oil. The water shines with the rainbow lines of oil. Oil boats loaded with oil, throng the oily stream, and oily men with oily hands fasten oily ropes around snubbing posts." [5] The muddy streets and the oily atmosphere, however, were not peculiar to Oil City, for these were characteristics of all the early oil towns.

Although it subsequently suffered from disastrous fires and floods, and was handicapped by the narrowness of the flat on which it was situated, Oil City continued to thrive. With the extension of the Atlantic & Great Western Railroad, the laying of pipe lines, the building of iron storage tanks, and the advantage of water transportation, Oil City emerged at the end of the decade as a prosperous and important shipping point for an extensive producing district; but it was destined to play even a greater role in succeeding years. Moreover, during the next decade, Oil City had more time for the amenities of life, and it followed the pattern of Titusville in its cultural development.

Meanwhile Titusville, possessing several advantages over

[4] *Oil City Register*, March 30, 1865.
[5] Bone, *Petroleum and Petroleum Wells*, 80.

Oil City and the other oil towns, rapidly developed. Located in a broad valley with room for expansion, it was already a village of several hundred people with certain social institutions when Drake struck oil. Being the nearest town to the site of the Drake well, Titusville naturally became the immediate destination of all who rushed to Oil Creek, and quickly assumed a more prominent position than any other place in the region. The completion of the Oil Creek Railroad from Corry to Titusville in the fall of 1862 increased its commercial importance. With a direct railroad connection at Corry to the east and the west, Titusville became not only an oil shipping center, but the most convenient approach to the oil field for thousands of people. Then the rise of the speculative boom, the completion of the Frazier well, and the spectacular events at Pithole had a most exhilarating effect upon the town.

By the summer of 1865, as a result of these influences, Titusville had developed into a bustling city of 10,000 population.[6] People thronged the streets; hotels and boarding houses were filled to overflowing, and 200 to 300 teams constantly hauled oil from Pithole to the numerous shippers' platforms on the railroad tracks, the receipts averaging over 1,000 barrels a day.[7] "In every highway and byway," one visitor wrote, "carpenters are at work and throughout the length and breadth of town new buildings are going up with magical rapidity."[8] It was estimated that over 500 new buildings were erected during 1865. Little attention was given to permanence of construction; the chief object was to build as quickly and as cheaply as possible. Businesses of all kinds had been established, and every one flourished: two dozen or more livery stables provided saddle horses to ride to Pithole; three banks with an aggregate capital of $750,000 fur-

---

[6] Titusville *Morning Herald*, Oct. 19, 1865, and March 2, 1866.

[7] *Ibid.*, Oct. 23, 1865.

[8] Correspondent of the *Pittsburgh Chronicle* quoted in the Titusville *Morning Herald*, June 27, 1865.

nished banking facilities; ten refineries had been built, and their weekly product amounted to 1,800 barrels; three telegraph offices kept busy day and night; the receipts of the express office amounted to about $30,000 a month; the sales of four hardware stores exceeded $75,000 a month; and there were numerous barrel factories, machine shops, and iron foundries, which manufactured everything for putting down a well.[9] Several of the largest grocery stores each did a business amounting to about $500,000.[10] The postal business had increased 600 per cent in a year, and the post office now ranked fourth in the state.[11] The trend was clear. The trade centering in Titusville, already much larger than at any other place along Oil Creek, grew livelier every day. Titusville was becoming the commercial and financial metropolis of the region.

Impressed by the rapid growth of Titusville and the development of the petroleum business, W. W. Bloss and H. C. Bloss came to Titusville in the spring of 1865, bought the *Petroleum Reporter,* which had been published intermittently as a weekly, and established the Titusville *Morning Herald,* the first daily newspaper in the oil region.[12] When it was announced that the Bloss brothers would publish a daily newspaper, wise men shook their heads and leading business men discouraged it. The publishers were not the kind to be discouraged, however, and went ahead with their plans. On June 14, 1865, the first issue of the Titusville *Morning Herald* came off the press. Starting with fewer than 400 subscribers, the circulation doubled before the end of the second month; and, contrary to all expectations, it proved self-sustaining. From the outset it was an influential fac-

---

[9] Titusville *Morning Herald,* July 21, Aug. 5, 7, 16, Sept. 5, Oct. 23, and Nov. 10, 1865.

[10] *Ibid.,* Oct. 23, 1865.

[11] *Ibid.,* Aug. 10, 1865.

[12] *Ibid.,* Sept. 18 and 28, 1865, and June 14, 1866; Titusville *Daily Evening Press,* Sept. 28, 1872.

tor in promoting and developing the best interests of Titusville. Furthermore, in a day before oil exchanges were established, its accurate and reliable monthly oil reports proved
of paramount interest to oilmen everywhere.

From the beginning of the rush to Oil Creek, Titusville's
leading citizens struggled to keep the village a peaceful, orderly, and law-abiding community, a decent place in which
to live. One of the first social problems they had to face
was the rapidly increasing number of saloons. In fact, so many
saloons were opened that the citizens held a mass meeting
at the Universalist Church on February 15, 1861, and re-
solved to put a stop "peaceably if they could, forcibly if they
must," to the sale of liquor.[13] Ten citizens were appointed
to visit the saloons and demand that the bars be closed and
all liquor removed from the borough. Followed by a large
crowd, the committee visited all unlicensed houses, read the
resolutions, and intimated force would be used unless the
nuisances were abated. And they were!

As the tide of travel swelled during the summer of 1865,
Titusville became infested with gamblers, pickpockets, coun-
terfeiters, horse thieves, prostitutes, and all sorts of vicious
characters. With only a chief of police and two patrolmen,
it was difficult to enforce the law, and because of personal
insecurity the practice of carrying revolvers became quite
common among the citizens.[14] In the fall, an increasing
number of incendiary fires excited and aroused the whole
town. "It is time for action," the editor of the Titusville
Morning Herald vigorously declared, "for secret, vigilant,
thorough, and united organization. It is time that persons
loafing about this town, without occupation, and without
visible means of support, were placed under unsleeping and
systematic surveillance; their haunts pried into; their associates scrutinized, and their steps followed with an unfal-

13 Crawford Journal, Feb. 26, 1861.
14 Titusville Morning Herald, Oct. 7, 1865.

tering pursuit." [15] The editor even suggested that "Judge Lynch" might exercise a wholesome influence.

A crisis was reached on the night of January 21, 1866, when two incendiary fires simultaneously broke out. The moment for a moral as well as a physical purification of the city had arrived. Immediately the authorities took into custody Robert Vance, alias "Stonehouse Jack," and three companions. An excited crowd milled about the jail and would have hanged the four if an opportunity had offered itself. The next morning a public mass meeting was held at the Board of Trade rooms to take measures to protect the city.[16] At the appointed hour the rooms were filled to overflowing with substantial citizens. After considerable discussion, they appointed two committees: a Committee of Ten to warn all thieves, pimps, harlots, and others to leave town on the next train; and a Vigilance Committee, composed of thirty men, to inquire into the origin of the fires and take proper action. The mass meeting then adjourned in order that the Vigilance Committee might proceed with its investigation. While the committee deliberated, some of the indignant citizens erected a gallows on a vacant lot and prepared for action. Fifteen suspected persons were brought before the committee for examination, and while none of them were responsible for the fires their character and reputation were considered so inimical to the community's safety that the committee ordered their expulsion from the city. It also notified a large number of prostitutes to leave. The next day the entire lot was escorted to the train by several hundred citizens, and started for Corry.

As a result of the citizens' movement, Titusville became an incorporated city on April 2, 1866, which enabled it to enlarge the police force and exercise other necessary municipal powers. The vigilance of the police and the firm and resolute

[15] *Ibid.*, Oct. 16, 1865.
[16] *Ibid.*, Jan. 23 and 24, 1866.

policy of the mayor in suppressing various dens of iniquity soon worked a revolution which almost made the city a model of virtue and morality. The fact that law and order prevailed helped to induce oilmen to bring their families and make Titusville their permanent home.

With the collapse of the speculative boom in 1866, Titusville felt the pinch of hard times. Business became dull, travel moderate; and those who had rented shops, hotels, and houses for exorbitant prices in good times, now grumbled and complained. To make matters worse, the teamsters, who had filled the streets and supported all sorts of businesses, were fast disappearing on account of the introduction of pipe lines. Not only this was true, but the building of pipe lines from Pithole to Miller farm, Shaffer, Oleopolis, and Henry's Bend created small business centers which took trade away from Titusville. With the falling off of business, the editors of the Titusville *Morning Herald* considered suspending publication. Upon hearing the rumor, Major S. M. Mills, the genial host of the Moore House, invited them to board at his hotel free of charge; he is reported to have said, "If the *Herald* stopped, my God, the town would go plum dash to destruction." [17] Despite dull times, business seemed to be better in Titusville than at any other town in the oil region because it had developed an immense mercantile trade with the surrounding country and was not solely dependent upon oil.

Fortunately, the opening up of new oil territory at Shamburg and Pleasantville in 1868 greatly stimulated business, prosperous times returned to Titusville earlier than to any other place in the oil region. Like Pithole, it once again became the center of travel and the base of supplies for Pleasantville. In every branch of trade and industry there were unmistakable signs of increased prosperity. On account of the lack of accommodations at Pleasantville, the hotels in Titusville were once more taxed to capacity, and every

[17] *Ibid.,* June 19, 1872.

man who owned a horse reaped a golden harvest, for all day long and nearly all night, Sunday and week days, the whole city rushed to Pleasantville in the morning and back at night. The demand for oil reports and news from Pleasantville increased the daily circulation of the Titusville *Morning Herald* 75 per cent and enabled the Western Union Telegraph office in Titusville to rank next to Philadelphia and Pittsburgh in the amount of business transacted.[18] The passenger and freight traffic of the Oil Creek Railroad to and from Titusville increased with wonderful rapidity. The regular passenger trains carried extra coaches, invariably crowded; ticket sales at Titusville during the summer jumped from $600 to $1,000 a week![19] Extra freights, loaded to capacity, ran in both directions.

By the end of the decade, Titusville emerged as the commercial and financial capital of the oil region. It was far ahead of other oil towns in the industries connected with oil production—boilers, engines, drilling tools, all kinds of machinery, barrels, hardware, and lumber.[20] For the nine months ending December 1, 1869, the amount of business transacted by a few of the representative houses was as follows:[21]

W. H. Abbott, petroleum	$458,955
Stewart and Stewart, refiners	454,000
Bryan, Dillingham & Co., ironworks	444,161
Granger & Company, grocers	373,285
Clark, Hayes & Company, grocers	354,500
Gibbs, Russell & Co., ironworks	286,500
McEowen & Company, grocers (3 months)	202,962
F. W. Ames, hardware	185,624
Murray & Company, refiners	160,000
Hinkley & Allen, refiners	120,000
Morland & Company, refiners	100,000
Jackson & Cluley	100,000

[18] *Ibid.*, Oct. 16, 1868, and Feb. 3, 1869.
[19] *Ibid.*, Aug. 8, 1868.
[20] *Oil City Times* quoted in the *Titusville Morning Herald*, Feb. 21, 1870.
[21] *Titusville Morning Herald*, Jan. 29, 1870.

There were four dry-goods stores whose individual business amounted to $100,000 to $125,000 a year. Deposits in the two banks exceeded $1,000,000, while the transactions at the banks averaged $75,000 a day. There was scarcely a business firm whose transactions did not exceed $25,000. Titusville was a great entrepôt.

As the city grew, certain changes in its appearance were evident. In 1865, logs furnished street crossings for the citizens, and in bad weather the road gave one a vivid idea of riding over a chain of mountains. By 1870, however, most of the streets had been graded and drained, and a large proportion of them improved. New plank walks extended in every direction, street lamps lighted by gas had been located all over town, and ground had been set off for a public park.

With the firm establishment of Titusville as the great emporium of the oil region, oilmen and others decided that it might be possible to go farther and do worse than make their home in Titusville. The result was an active demand for lots, and a building boom ensued. During the summer of 1870 over 300 dwellings were erected, representing an investment of about $1,025,900.[22] In addition to the smaller houses, many large and stately residences, costing from $10,000 to $60,000, were built. Titusville had never seen a year in which such extensive investments in buildings had been made.

The absorption in material things and the eagerness to acquire wealth did not cause the citizens of Titusville to forget the educational, cultural, and spiritual side of life. Long before Colonel Drake struck oil, the Presbyterians, Roman Catholics, Methodists, Universalists, and Baptists had organized societies, erected churches, and held services. With the rush to Titusville and Oil Creek, the Episcopalians organized a society in 1862 and built a fine church in 1864.

[22] *Ibid.*, Sept. 30, 1870.

Taking advantage of the prosperous times, all the other denominations began constructing new churches within the next eighteen months.

A splendid common school, a soldiers' orphans' school, and a Roman Catholic school had been established; but there was a belief that Titusville should have something beyond the grades, and so a group of the leading citizens founded Appleton's Collegiate Institute. It opened in the fall of 1865 under the supervision of a committee of the vestry of the Episcopal Church, but it was not denominational in its teaching.

In July, 1864, under the leadership of John D. Archbold and a few others, a group of men organized the Culver Literary Association, furnished a reading room, and purchased books. When it was reorganized in December, 1865, as the Titusville Literary Association, the meetings were well attended and the membership increased to 170. It was probably this group which brought Artemus Ward to Titusville in November, 1865, for the first public lecture in town.

The enthusiastic response on the part of the public to Ward's lecture led to the appearance of many distinguished lecturers, who either came on their own initiative or were sponsored by the Young Men's Christian Association. Between 1865 and 1871 Charles Sumner, Theodore Tilton, Josh Billings, S. M. Hewlett, Olive Logan, A. A. Willetts, Frederick Douglass, John B. Gough, William Parsons, Wendell Phillips, Kate Field, P. T. Barnum, Mark Twain, and Horace Greeley visited Titusville and spoke before capacity audiences. Some, like B. F. Taylor, Isaac I. Hayes, Anna E. Dickinson, Bishop Simpson, and General Judson Kilpatrick, lectured two, three, and even four different times.

Fully as brilliant as the talent appearing on the lecture platform were the musicians who gave concerts. When it was announced that Clara Louise Kellogg, distinguished American prima donna, would sing in Titusville, tickets for seats in the parquet, dress circle, and family circle were sold out days in advance.[23] On the day of the concert, people came from Oil City, Tidioute, Corry, Franklin, and surrounding towns, and that evening every seat from orchestra to gallery was filled. It was a scene of great animation and brilliancy. The "petroleum aristocracy" from all over the region had assembled, and the women presented a fashionable display of dresses, hats, shawls, cloaks, diamonds, silks, and laces. It was the first important musical as well as social event in the city. An equally fine audience greeted the Richings-Bernard English Opera Combination, which presented the first opera in Titusville, Theodore Thomas and his orchestra, and James Fisk's Opera Bouffe and Ballet Company of New York. Probably the greatest musical treat of the decade was the concert by Christine Nilsson, the eminent Swedish operatic soprano.

Beginning in 1865 many dramatic companies played in Titusville and were well patronized. Susan Denin, Eloise Brydges, Evelyn Evans, Julia Day, Kate Ryner, and N. D. Jones of the Howard Athenaeum in Boston appeared, playing in "East Lynne," "The French Spy," "Camille," "Lucretia Borgia," "The Octoroon," "Our American Cousin," "Othello," "Hamlet," "Richelieu," and others. James F. Sherry's New York Theatre, including J. W. Carner, S. K. Chester, and Virginia Howard, opened the new Parshall Opera House in December, 1870, in "Rip Van Winkle," and continued for several weeks, playing to capacity houses. The most brilliant dramatic event, however, was the appearance of the great German tragedienne Fanny Janauschek.

With the acquisition of wealth and leisure, some of the affluent citizens went in for horse-racing. They organized the Titusville Driving Park Association in 1870, leased a fifteen-acre plot west of town, and laid out one of the best half-mile tracks in the region. The June races were gala

23 *Ibid.,* Jan. 6, 1871.

days; the whole town was full of excitement, and the streets were crowded. The races were no ordinary affairs because valuable horses had been purchased by prominent citizens for racing purposes. Among the fast horses was "Acuff," owned by Jonathan Watson, R. D. Fletcher, and A. H. Carr and valued at $5,000; "McDuff," owned by Ezra Crossman and valued at $5,000; "Lillie," R. D. Fletcher's brown mare; and D. H. Cady's "Ben Wade." [24] The average value of the horses owned in Titusville ranged from $2,000 to $3,000. Not only did Titusville have some of the fastest horses in the region, but it had a reputation for having the best and most honorably conducted races.

In view of all these distinguishing characteristics, which made Titusville different from other towns in the oil region, one might readily agree with the local editor, who referred to the Queen City as a "city of taste, wealth, refinement and culture, with half a score of churches, its schools, newspapers, markets, elegant stores, lecture associations, social clubs, and daily supplies of oil and everything else needed." [25] Titusville was not only a commercial center of importance but an attractive place in which to live. To ride into Titusville, wrote a correspondent of the *New York Tribune,* was an "inexpressible relief, for Titusville though an oil city, getting its wealth and owing even its existence to petroleum, is so far off on the verge of the flowing district that it spares some time to make itself decent, and really look as if it intended to last. . . . Here speculators and operators congregate, here the statistics and news of the whole Venango region are collected and discussed; here shippers transact a great part of their business, and the money-getters, who make their life among the wells, run hither every day or so and refresh themselves with a glimpse at civilization." [26]

---

[24] *Ibid.,* March 20 and Aug. 8, 1871.
[25] *Ibid.,* Oct. 14, 1871.
[26] Quoted in the Titusville *Morning Herald,* Aug. 11, 1868.

CHAPTER XIV

Oil Rings and Oil Exchanges

WITH the growth of the petroleum industry and a concentration of much of the buying and selling at Oil City and Titusville, it was only natural that these two cities should become the home of the first organized oil exchanges. Early buyers, or their agents, generally traveled through the oil field on horseback, buying oil here and there in any quantity and making the best bargain possible with each individual producer. As a matter of fact, sales might be made anywhere. Buyers and sellers often gathered in groups on the streets of Titusville or Oil City, in hotel lobbies, or in front of the telegraph offices, and bought and sold petroleum according to the best information received during the day. Gradually the oilmen from upper Oil Creek gravitated to the American Hotel in Titusville, kept by the well known and popular Major Mills, the father-in-law of John D. Archbold, where much of the buying and selling took place. Here the Atlantic and Pacific Telegraph Company had an office, market reports could easily be obtained, and the American Hotel literally became the first oil exchange. Later, producers, dealers, and speculators at Oil City congregated on the sidewalk in front of the office of Lockhart, Frew and Company, near the railroad track, discussed new strikes, told stories, and bought or sold oil, forming the "Curbstone Exchange." Frequently, the business transacted here each day amounted to $500,000 or more.[1]

[1] *Titusville Morning Herald,* May 15, 1871.

Written contracts for sales were unnecessary in the early days, for an oilman's word was as good as his bond. Illustrative of the manner in which business was done and the evidence of good faith among early oilmen is the story of Charles L. Wheeler of Titusville and Orange Noble.[2] Soon after the completion of the Noble and Delamater well, Wheeler met Noble one day on the streets of Titusville and asked him what he would take for 30,000 barrels of oil. Noble asked $1.50. Wheeler said he would take it, but they did not make any written record of the transaction. Before the oil was delivered, the price rose to $7.50 a barrel, but despite the rise Noble delivered every barrel for $1.50 just as if a formal contract had been executed.

Most oil dealers conducted their business as a series of speculations and wagers as to what the price might be on a certain day in the future. Price changes were wide, sudden, and violent, shifting all the way from $20 a barrel in 1859 to ten cents a barrel in the fall of 1861. In March, 1862, the prices ranged between twenty and twenty-five cents a barrel; but in 1863 a decline in production, an increase in consumption, and improved facilities for handling oil caused the price to rise to $4. This uncertainty in price and demand made petroleum the favorite speculative commodity of the time. The bulls bet on higher prices and labored to produce them, while the bears bet on lower prices and did everything possible to depress them. There were constant attempts by different parties to force the price of oil to a higher figure by a certain date by obtaining complete or partial control of the oil in the district, forcing those who had sold oil for future delivery to settle at advanced prices. It was a wild game, which brought either riches or ruin.

In 1869 oilmen witnessed for the first time two great bull movements which greatly influenced prices, and about a dozen others of lesser importance. All together the market

[2] "Charles L. Wheeler," *Petroleum Age*, IV (1888), 1184–1185.

was under the control of the speculators for about nine months of the year, a consistent effort being made for six months to advance prices and for three months to depress them.[3]

The first large combination, without parallel in the oil trade, originated in New York in 1868 and had members throughout the United States and Europe.[4] Apparently the leading bankers of Europe financed the affair and aimed at controlling the export trade. When the ring first entered the market in December, 1868, the price of oil along Oil Creek was $4 to $4.25 a barrel; there was a steady upward swing in prices owing to heavy buying, and in February the price of oil reached $7, the highest figure for the year. The purchases of the ring were enormous; some were in lots varying all the way from 25,000 to 200,000 barrels; there was one purchase of 500,000 barrels for a Philadelphia refiner. The combination then allowed prices to decline gradually until oil reached $5.30 in May on Oil Creek, when the ring almost wholly withdrew from the market.

The second large combination occurred in August. Led by one of the shrewdest operators on Oil Creek, the bulls allied themselves with Pittsburgh refiners for the purpose of advancing the price of refined oil and creating a larger margin between it and the price of crude.[5] As it was, the difference was so small that few refineries, except those actually in the region, could be worked at a profit. Pittsburgh refiners agreed to stop business and create a scarcity in the Philadelphia market, to which nearly all of their product was sent. To shut off the supply for New York, the cooperation of the Cleveland refiners, the principal source, was sought; but they refused, either believing that the scheme would not be successful or else not being in a position to close their

[3] *Titusville Morning Herald*, Jan. 24, 1870.
[4] *Ibid.*
[5] *Ibid.*, Sept. 13, 1869.

THE DRAKE MONUMENT, TITUSVILLE

Erected in 1901 to the memory of E. L. Drake by Henry H. Rogers

Nelson Studio

MUSEUM   CARETAKER'S HOUSE   SITE OF DRAKE WELL

THE DRAKE WELL MEMORIAL PARK

To mark the site of the Drake well, this park was constructed and dedicated in 1934.

works. Despite the attitude of the Cleveland refiners, the engineers of the movement worked feverishly and seemed to gather strength. On September 1, the time agreed upon to shut down, about thirteen of the largest refineries in Pittsburgh were closed. The bulls began purchasing in small amounts, refined in New York and Philadelphia rose 1 to 1½ cents a gallon, while crude along Oil Creek became strong and rose 20 cents a barrel. Though no great appreciation of prices occurred, the ring created something of a scarcity in refined oil. The bulls continued their activity, which became the topic of conversation all over the oil field. They purchased 50,000 barrels outright, and an equal amount at seller's option in October, Philadelphia delivery.[6] With the foreign demand at this time of year exceptionally large, the bull movement caused a stiffening of prices in view of a scarcity of oil. During the first two weeks of October, the bulls actually held out of the market 100,000 barrels of oil, and then about the 19th called on buyer's options which they owned.[7] This caused a commotion among those who expected to procure oil to settle contracts, as it created a corner on oil on the last two days of the month. After the 19th, the price steadily advanced, and it stood at 39 cents on the 29th, the date contracts were due. Most October contracts were settled on the basis of 36 cents. How badly the shorts were squeezed, no one knew; but some lost heavily. Outside Philadelphia and Pittsburgh, the markets were only slightly affected by the corner.

The most famous oil corner originated in Pittsburgh almost simultaneously with Gould and Fisk's attempt to corner the gold supply in 1869.[8] Although others were

[6] *Ibid.*, Nov. 1, 1869. Oil purchased according to seller's option could be delivered any time within the period specified by the contract, provided the buyer had been given ten days' notice.

[7] *Ibid.* Oil purchased according to buyer's option could be called for delivery any time within the period specified by the contract, provided the seller had been given ten days' notice.

[8] *Ibid.*, Oct. 4, 1871, Nov. 9 and 18, 1869.

involved in the oil ring, Lockhart and Frew, Alex. Byers, Joseph Fleming, David Hostetter, Warden, Frew and Company, and the Allegheny Valley Railroad Company engineered the movement. Of the men involved, one writer said: "The conspirators are architects of their own or of others' ruin. And yet they are all honorable men, and sat at the head of their church pews yesterday, and some of them perhaps raised the communion cup to their lips, and bowed to the prayer, that they might be enabled to do unto others as they would that others should do unto them." [9]

Long before unwary oil dealers discovered that a movement was under way to corner the market, the ring had quietly and without suspicion purchased many contracts for the delivery of oil in December. With $1,000,000 or more in the pool, the combine expected to buy all the oil in the market, and then sell later to those who had contracts to fill at the close of the year at whatever price the ring decided. For a while the plan prospered. The ring purchased contract after contract, and as they fell due the oil was delivered. According to custom, when a contract came due and the maker could not deliver the oil, he paid for the oil at market price, whatever it might be. But in this instance the ring refused to accept anything but oil. Even if the sellers offered 20 cents a gallon more than the market price, the ring refused: oil it must and would have. Persons with contracts due were compelled, therefore, to buy from those who had no contracts, or from those whose contracts were not due until later.

To make the situation more favorable for the Pittsburgh ring, the Allegheny River fell to a low stage and remained there, so that no oil could be shipped. The Allegheny Valley Railroad, being a part of the combination, was always ready with the excuse that it had no cars in which to ship oil

[9] *Ibid.*, Nov. 9, 1869.

to Pittsburgh.[10] As October and November deliveries fell due, oil grew scarce, and prospects looked gloomy for a large number with contracts to deliver oil at the close of the year. Fortunately, the river had a rise in December, and oil which had accumulated in large quantities above came down to take advantage of the high prices, thus making it possible for the shorts to fill contracts. The ring, in order to save itself, entered the market and bought oil to exhaust the supply and keep up prices until December 31. It was all in vain, however, as the Allegheny River continued to be navigable; and despite the activities of the ring all who had contracts for the delivery of crude oil on December 31 were able to fulfill them at a price of 18 cents a gallon. The next day the price dropped to 12, leaving the ring in possession of over 500,000 barrels that had cost on an average about 15 cents. With the supply on hand, there was little prospect for a rise in price, and so the combination had to make the best of the bargain. The bulls had been disastrously defeated, the ring exposed, and for weeks it was the most exciting topic of conversation in the oil region. Members of the ring came in for considerable censure because nearly all the money available for loan purposes in Pittsburgh had been utilized by the combine, causing a shortage for use in manufacturing and mercantile enterprises.[11] As in this case, corners rarely resulted in profit to the originators. Generally, some petroleum dealer opened his tanks, broke the corner, and helped out the cornered individuals by furnishing oil.

Dealers in petroleum not only were victimized by intriguing cliques, which ran the price up and down by devious methods, but suffered from other handicaps in marketing oil. They lacked correct information regarding the amount of oil produced, a necessary factor in determining the price. Moreover, buyers seldom quoted the same price to producers

[10] *Ibid.*, Oct. 4, 1871.
[11] *Ibid.*, Dec. 28, 1869.

at widely separated points. The ridiculousness of the situation is indicated by the fact that on a certain day in October, 1865, the quotation on 1,000 barrels of oil at Petroleum Centre was $7.25 a barrel, at Bennehoff Run $8, and at Bull Run $9.[12] It frequently happened that a shipper, receiving word of an important occurrence in New York, would mount his horse and start either up or down the Creek, buying all the oil possible. Within a short time, he was followed by some one willing to pay 25 cents more a barrel. The second might be followed by a third who was willing to go still higher. On the other hand, conditions might be reversed so that shippers would be compelled to sell for less than the purchase price. No one seemed to know or care about actual market conditions, each selling or buying to suit his own particular fancy. A third handicap in ascertaining the real market value of oil was the fact that most transactions in the oil region were not known; as they were not a matter of record, no correct quotation could be given. Under the circumstances there was a definite need for an organization where buyers and sellers could meet, ascertain with some accuracy the amount of oil actually being produced and sold, and transact their business in accordance with certain rules and regulations.

Some of the influential oilmen of Titusville attempted to establish an oil exchange in 1864, but without success.[13] In November, 1865, however, the merchants and oil shippers organized a Board of Trade, leased commodious rooms in Waite's new brick building, had a telegraph line installed, elected J. H. Bunting as president, appointed a standing committee on arbitration, set the hours from 1:30 to 3:30 P.M. for trading, and on November 21' held the first meeting.[14] This was the first organization of its kind in the oil region.

12 *Ibid.*, Oct. 21, 1865.
13 *Ibid.*, Oct. 24, 1865.
14 *Ibid.*, Nov. 10, 11, 18, 20, 22, 1865, and Feb. 10, 1866.

The rooms were open at all hours of the day; a dozen or more of the principal daily papers were on file, and telegraphic reports from the principal markets were received. Even though the mercantile interests of Titusville were included, the producers and buyers were at the outset so well represented that it functioned primarily as an oil exchange.[15] After a three months' trial, it was found that the buyers attended the meetings of the Board of Trade, but few sellers were present.[16] Therefore, the shippers were forced to continue their traveling over the oil fields offering high prices here, low there.

When the railroad was completed between Oil City and Titusville in 1866, many of the passengers were oilmen. The road was rough, the stops numerous, and the seventeen-mile trip required three hours. As an accommodation, the railroad put on a passenger car especially for oilmen, and as they rode along they bought and sold oil.[17] In reality the passenger car constituted an oil exchange, although it had no officers or rules regulating the conduct of business.

As the excitement and productivity increased at Petroleum Centre, producers and dealers in that vicinity agitated for the organization of an oil exchange. In October, 1867, thirty-four of them, with Edward Fox as president, organized the Oil Dealers' Exchange, which, being central and easily accessible, was expected to have 100 members within a relatively short time.[18] Until a suitable building could be secured, the meetings were held in the Opera House, where they received daily telegraphic reports from eastern petroleum markets. Though the depression made business dull, the Oil Dealers' Exchange received a fair amount of support among the oilmen.

[15] *Ibid.*, Jan. 6, 1866.
[16] *Ibid.*, April 10, 1866.
[17] "Speculative Halls—The Oil City Oil Exchange," *Petroleum Age*, IV (1885), 975–980.
[18] Titusville *Morning Herald*, Oct. 5, 22, 23, 24, and Nov. 19, 1867.

In February, 1869, the producers met at Oil City to consider methods of protecting their interests and correcting abuses in the trade. At the next meeting in Titusville, they organized the Petroleum Producers' Association of Pennsylvania and became very active at once in opposing a bill in the State Legislature to tax crude oil, in preparing to issue monthly reports on production and operations, and in attacking the torpedo monopoly.[19]

During the same month, the oilmen along upper Oil Creek held a meeting at the American Hotel in Titusville to discuss the establishment of an exchange, but nothing definite resulted.[20] Finally, in January, 1871, one hundred and sixteen dealers, producers, refiners, and brokers along Oil Creek agreed to form an association to be known as the Titusville Oil Exchange and abide by such rules as might be adopted.[21] During the next two months they perfected the organization, secured new members, drafted rules and regulations, and selected and beautifully furnished rooms in the Parshall House. On March 13, the members of the Titusville Oil Exchange met for the first time in their new rooms and commenced business.[22]

From the beginning, oilmen crowded the Oil Exchange, especially in the evenings, and it was with difficulty that one could pass through the crowd. In summer weather, the sidewalk in front presented a lively scene. Within a year the new institution had a membership of 150, including refiners, dealers, brokers, and representatives of every branch of the oil trade. Located in the heart of oildom and in the home town of many operators, brokers, and others connected

[19] *Derrick's Hand-Book*, I, 111, 113, 116; *Petroleum Centre Daily Record*, Feb. 24, 1869.

[20] *Derrick's Hand-Book*, I, 111; Titusville *Morning Herald*, Jan. 27 and Feb. 6, 1869.

[21] Minutes of the Titusville Oil Exchange, 1; *Titusville Morning Herald*, Jan. 14, 16, 18, and 24, 1871. The by-laws and articles of the organization are printed in the *Titusville Morning Herald*, March 16, 1871.

[22] Minutes of the Titusville Oil Exchange, 9.

with the industry, the Titusville Oil Exchange flourished and became an important influence in the marketing of oil.

Within a month after the establishment of the Titusville Oil Exchange, another was organized at Franklin.[23] Since the lack of a suitable place in which to transact business had been long felt at Oil City, the oilmen located here organized an exchange in May, 1871, secured rooms in the Sands block, moved to the Opera House block in 1872, but disbanded when complications arose in 1873 over the Southern Improvement Company.[24] For over a year, oil transactions were made on the streets, in offices, and at hotels. Finally, the exchange was reorganized in 1874, with headquarters in the Collins House. The Oil City Oil Exchange quickly assumed the leadership in buying and selling, and occupied a commanding position for many years.

These three—at Titusville, Franklin, and Oil City—constituted the pioneer oil exchanges, and they were destined to play a spectacular role in a rapidly growing industry. As the developments spread, other exchanges, patterned after these, were later established at Pittsburgh, Bradford, Parker, Petrolia, and other oil towns.

[23] The oil exchange in Franklin was organized in April, 1871 (*Venango Spectator*, April 21, 1871).

[24] *Titusville Morning Herald*, May 15, 1871; "The Speculative Halls—The Oil City Oil Exchange," *Petroleum Age*, IV (1885), 975–980.

CHAPTER XV

The End of a Decade

OIL CREEK and its adjacent fields continued to be the largest oil producing district in 1869; but with the decline in production at Shamburg and Pleasantville, the principal scene of operations now shifted down the Allegheny to Butler, Armstrong, and Clarion counties. Operators, speculators, and all others who were accustomed to following the excitement upon the opening of a new field, rushed to Parker's Landing, Argyle, Petrolia, Karns City, and other places. As the leadership in production passed to the district below Franklin in 1873, thence up the Allegheny to Warren and Bradford in 1878, the Oil Creek region could content itself with the thought that it had given birth to the oil industry and witnessed its emergence as one of the greatest and most important in the nation.

The rise and growth of the industry had been truly remarkable. By 1870 it had grown until no less than $200,000,-000 was invested in the business.[1] From 2,000 barrels in 1859, the annual production of the Pennsylvania oil region jumped to over 4,800,000 in 1869, and the daily production averaged over 13,000 barrels.[2] The tremendous increase in production had been accompanied by an increase in consumption, so that the supply and demand in 1869 were more nearly equal than during any preceding year. Like-

[1] Ida M. Tarbell, *The History of the Standard Oil Company*, I, 36.
[2] *Titusville Morning Herald*, Jan. 24, 1870; *Derrick's Hand-Book*, I, 120.

wise, the expansion of the export trade in petroleum had been phenomenal. In 1859 we did not export any petroleum; ten years later, we were exporting over 96,000,000 gallons a year, valued at more than $30,000,000.[3] Most striking was the fact that petroleum now ranked second in value only to cotton among our exports! The refining business had also assumed large proportions. The refineries in Titusville had a capacity of 2,856 barrels a day; those in Pittsburgh refined 20,100 barrels a week, but their capacity was four times this amount; those in New York had about the same capacity as those in Pittsburgh; and Cleveland had numerous refineries with a daily capacity of about 16,000 barrels.[4]

The new industry not only had an unusual growth, but exerted a marked influence upon the world. Coming into existence at a time when the problem of cheap artificial light was becoming serious, the principal refined product, kerosene, had everywhere displaced all other means of illumination except gas. It was cheap, efficient, and safe. Of its social effect and universal usage, one authority wrote: "It is the advent of refined petroleum at comparatively low prices that has practically lengthened the duration of human life and has added vastly to the social enjoyment of mankind, not only among highly civilized peoples, but among the semi-civilized and barbarous nations; in fact, wherever the white wings of commerce can transport it there it has gone, and more, its light has penetrated even the solitudes of eastern deserts and the forests of both hemispheres." [5]

Though used principally for illuminating purposes, petroleum was also beginning to be utilized in other ways. Prior

[3] *Annual Report of the Deputy Special Commissioner of the Revenue in Charge of the Bureau of Statistics, on the Commerce and Navigation of the United States, for the Fiscal Year Ended June 30, 1869*, part I, 222–223.

[4] *Derrick's Hand-Book*, I, 140; *Titusville Morning Herald*, Dec. 19, 1870, May 18, 1872.

[5] Peckham, "Production, Technology, and Uses of Petroleum and Its Products," *House Misc. Doc.*, No. 42, 47 Cong., 2 sess., part 10, 261.

to 1869, its offensive odor prevented its extensive use as a lubricant; but in that year Joshua Merrill of the Downer Company discovered a process by which a deodorized lubricating oil could be manufactured. It quickly found a ready market and contributed substantially to the birth of the modern industrial era. Petroleum was also used as a substitute for turpentine in the manufacture of varnish, mixing paints, and for other purposes for which turpentine was essential. Its use as a fuel was being discussed and investigated. In 1866 it was reported that the British navy tried to burn oil in some of its vessels with satisfactory results.[6] On the other hand, the United States navy in 1867 conducted a complete series of experiments with oil as a fuel, but the results were unsatisfactory.[7] The apparatus for burning was simple and safe, but there were several serious objections to its use: the oil might explode as a result of a single shot piercing the ship; the loss of oil by volatilization was very great, thereby increasing the cost; and the odor was intolerable. "From these considerations it appears," the chief of the Bureau of Steam Navigation concluded, "that the use of petroleum as a fuel for steamers is hopeless; convenience is against it, comfort is against it, health is against it, economy is against it, and safety is against it. Opposed to these the advantages of the probably not very important reduction in bulk and weight, with their attending economies, cannot prevail." [8]

Finally, the discovery of petroleum in large quantities not only provided a new and cheap illuminant at a time when one was badly needed, but proved of considerable value to the Union during the Civil War. At the beginning, the North was cut off from the turpentine districts of the South and the small supply on hand, not being adequate for painting

[6] *Derrick's Hand-Book*, I, 73.

[7] "Report of the Secretary of the Navy," *House Exec. Doc.* No. 1, 40 Cong., 2 sess., 173–178.

[8] *Ibid.*, 175.

and other purposes, brought exorbitant prices.[9] A satisfactory substitute for turpentine was derived, however, from
some of the products discovered in refined petroleum. Petroleum aided in another way; it helped fill the treasury of
the federal government in the greatest crisis of its history.
Under the law of July 1, 1862, and that of June 30, 1864,
levying a tax on refined and crude petroleum, the United
States collected close to $7,000,000 during the Civil War.[10]
The government also derived considerable revenue from a
newly created occupational group—the oilmen—under the
income-tax laws. Even when the war was over, the revenue
from petroleum aided in paying off a portion of the national
debt. Sir S. Morton Peto, the English banker, suggested
another manner in which petroleum proved useful to the
Union. He said: "At a moment of civil war, when the balance
of trade was against the nation, when gold was necessarily
going out, and when there was a heavy drain upon the natural
resources of the country, petroleum sprung up, from lands
previously valueless, in quantities sufficient to make a sensible
diversion in the national commerce." [11] Stating the matter a
little more strongly and pointedly, the editor of the Titusville
Morning Herald wrote, "Constituting a medium of exchange
with foreign countries, in the opinion of the importers,
petroleum alone enabled this country to successfully carry
on and terminate our Civil War." [12] When President Grant
visited Titusville in September, 1871, he spoke briefly to
the assembled throng in front of the Parshall House, and
emphasized the same point.[13] Nor were these all the benefits.
While petroleum became more important in our foreign

[9] Ise, *The United States Oil Policy,* 9.
[10] Hayes, "Report of the United States Revenue Commission on Petroleum
as a Source of National Revenue," *House Exec. Doc.* No. 51, 39 Cong., 1 sess.,
3, 33.
[11] Peto, *Resources and Prospects of America, Ascertained During a Visit to
the States in the Autumn of 1865,* 205.
[12] Titusville *Morning Herald,* July 27, 1866.
[13] *Ibid.,* September 13, 14, and 15, 1871.

trade, it stimulated the internal industry of the United States; it gave the railroads of the country a prospect of large additional profit; it offered employment to capital with every prospect of abundant returns; and it afforded a more than ordinary reward for labor. "It is difficult," Peto remarked, "to find a parallel to such a blessing bestowed upon a nation in the hour of her direst necessity." [14]

[14] Peto, *op. cit.,* 206.

Bibliography

I. Materials in the Drake Museum

THE Drake Museum at Titusville, Pennsylvania, is a storehouse full of all sorts of historical records and relics pertaining to the early history of the petroleum industry. Edwin C. Bell's famous collection forms the nucleus to which a variety of research materials has been added since 1934. All of the materials have recently been catalogued, and a card index is now maintained in the library; but no printed list of records or relics is available. While a complete description of everything in the Drake Museum is not possible here, some of the more important items or collections will be indicated.

The largest and most important single collection of manuscripts is the Townsend collection of papers, letters, and records relating to the organization of the Seneca Oil Company and the drilling of the Drake well. The papers had been kept by James M. Townsend of New Haven, leading spirit in the Seneca Oil Company, and were given to the Drake Museum in 1934 by his nephew, H. H. Townshend of New Haven. The Minute Book, covering the period from March 19, 1858, to March 7, 1864, and the Ledger of the Seneca Oil Company, original leases and contracts, Drake's reports, letters of Drake, maps, Mr. James M. Townsend's account of the Seneca Oil Company and its activities at Titusville are among the more important papers in the collection.

Other valuable manuscripts in the Drake Museum are: Colonel Drake's story of the Drake well, written about 1870; the early records of the Oil City Oil Exchange and the Titusville Oil Exchange; a leaf from the Ledger of R. D. Fletcher, Titusville merchant, showing Drake's account; the contract between S. M. Kier of Pittsburgh and E. L. Drake, dated November 14, 1859; a prospectus of the United States Petroleum Company; and the hotel registers of the old Moore House and American Hotel at Titusville and the Bonta House at Pithole. The Museum also has stock certificates of old oil companies, oil company reports, early oil leases, account books, photographs of early oilmen, and a collection of maps showing the early presence of petroleum in the United States.

About 1860 John A. Mather established a picture gallery in Titusville and took photographs everywhere throughout the oil region. Through the efforts of Edwin C. Bell, the Drake Memorial Association secured the Mather negatives, numbering about eighteen hundred. Many of them have since deteriorated, but several hundred are still in good condition and are now at the Drake Museum, constituting the most interesting and the finest collection of contemporary views of the oil region to be found in the country.

Bound files of the *Morning Herald, Weekly Herald, Daily Courier, Weekly Courier, Evening Courier,* all published in Titusville, and the *Venango Spectator* of Franklin, an unbound file of the *Oil City Register,* a number of copies of the *Petroleum Centre Daily Record, Pithole Daily Record,* and a miscellaneous collection of old Titusville newspapers may be found in the Museum.

Among the books are reports of the United States and several state geological surveys, many histories of petroleum, now out of print and scarce, city directories of oil towns, and personal accounts of early days in the oil fields. Some of the books are listed below.

Important in showing the development of lamps is the collection assembled by the late P. C. Boyle of Oil City. It includes a few very old grease-burning lamps, many types of candlesticks, candle-burning lanterns, lamps for burning camphene and other burning fluids, and many types of kerosene lamps.

On display in the Museum are all sorts of old oil-well tools, a few of which came from the Drake well. Especially interesting are several miniature models of methods used to drill and pump early oil wells.

II. OTHER MATERIALS

A. *Manuscripts*

The hitherto unpublished and unused letters of Dr. F. B. Brewer have been most valuable in supplementing the Townsend collection. The letters are of particular importance in reconstructing the story of the origin and organization of the Pennsylvania Rock Oil Company of New York and the Pennsylvania Rock Oil Company of Connecticut. Most of them were written by Albert Crosby, George H. Bissell, Jonathan G. Eveleth, and Anson Sheldon.

B. Newspapers

The following newspapers located in towns in and on the edge of the oil region contain a mass of contemporary material:

> *Crawford Journal* (Meadville)
> *Forest Press* (Tionesta)
> *Oil City Register*
> *Petroleum Reporter* (Titusville)
> *Pithole Daily Record*
> *The Crawford Democrat* (Meadville)
> *The Bee* (Tionesta)
> *The Oil City Derrick*
> *The Petroleum Centre Daily Record*
> *The Venango Spectator* (Franklin)
> *The Warren Mail*
> *The Warren Ledger*
> *Titusville Gazette and Oil Creek Reporter*
> *Titusville Weekly Herald*
> *Titusville Morning Herald*

The *Oil City Derrick* for August 27, 1909, the Diamond Jubilee issue of the *Titusville Herald* of August 22, 1934, and the Diamond Jubilee issue of the *Oil and Gas Journal* of August 23, 1934, are especially valuable for the student of early oil history.

C. Periodical Literature

Many articles on the early days in the oil region may be found in various issues of *Harper's New Monthly Magazine,* the *Century Magazine,* the *Nation,* the *Atlantic Monthly,* the *Petroleum Age,* the *Lamp,* the *Texaco Star,* the *Scientific Monthly,* the *Petroleum Monthly,* the *Western Pennsylvania*

Historical Magazine, and other periodicals. A few of the more important are listed below.

Signed Articles:

CRUM, A. R., "A Pithole Legend of J. Wilkes Booth," *Petroleum Age,* VI (1887), 1733.

CAREY, CRONEIS, "Early History of Petroleum in North America," *Scientific Monthly,* XXXVII (1933), 124–133.

EATON, AMASA M., "A Visit to the Oil Regions of Pennsylvania in 1865," *Western Pennsylvania Historical Magazine,* XVIII (1935), 189–208.

FRANKLIN, B., "After Petroleum," *Harper's New Monthly Magazine,* XXI (1864), 52–64.

MURRAY, F. F., "Oil Men of Prehistoric Days," *Texaco Star,* Aug., 1930.

SCHOOLEY, JOHN S., "The Petroleum Region of America," *Harper's New Monthly Magazine,* XXI (1865), 562–574.

SMALLEY, E. V., "Striking Oil," *Century Magazine,* (1883), 323–339.

TROWBRIDGE, J. T., "A Carpet-Bagger in Pennsylvania," *Atlantic Monthly,* XXIII (1869), 729–747.

Unsigned Articles:

"Coal Oil Johnny," *Petroleum Age,* VII (1888), 16.

"Col. Drake," *Petroleum Monthly,* I (1870), 1–9.

"George M. Mowbray," *Petroleum Monthly,* I (1871), 117–120.

"Oil Creek in 1865," *Petroleum Monthly,* I (1871), 194–200.

"Pithole," *Nation,* I (1865), 370–372.

"Speculative Halls—The Oil City Oil Exchange," *Petroleum Age,* IV (1885), 975–980.

" 'Spirit' and Other Influences in Locating Wells," *Petroleum Age* I (1882), 91–92.

D. Books

The best general histories covering the early development of the petroleum industry are: *The Early and Later History of Petroleum, with Authentic Facts in Regard to Its Development in Western Pennsylvania,* by J. T. Henry (Philadelphia, 1873), and *The History and Romance of the Pe-*

troleum Industry, by James Dodd Henry (London, 1914). A more popular account is *Sketches in Crude-Oil* (Harrisburg, 1896) by John J. McLaurin. Much of the material on oil in the *History of Venango County, Pennsylvania, and Incidentally of Petroleum, Together with Accounts of the Early Settlement and Progress of Each Township, Borough, and Village, with Personal and Biographical Sketches of the Early Settlers, Representative Men, Family Records, etc.,* edited by J. H. Newton (Columbus, 1879), consists of direct quotations from earlier writers. An excellent but brief account of the early history of petroleum may be found in *The History of the Standard Oil Company,* by Ida M. Tarbell (New York, 1904). *The Romance of American Petroleum and Gas,* by A. R. Crum and A. S. Dunagan (New York, 1911), contains considerable historical material about the early developments. *The Derrick's Hand-Book of Petroleum: A Complete Chronological and Statistical Review of Petroleum Developments from 1859 to 1899* (2 vols., Oil City, 1898) is an indispensable mine of information. F. W. Beers' *Atlas of the Oil Region of Pennsylvania* (New York, 1865) is an important aid for locating leases, farms, roads, wells, towns, creeks, and railroads in the oil region. *Petroleum in the United States and Its Possessions,* by Ralph Arnold and William J. Kemnitzer (New York, 1931), is a valuable reference work for data on the early oil fields.

Excellent contemporary accounts are: *The Oil Regions of Pennsylvania: Showing Where Petroleum Is Found; How It Is Obtained, and at What Cost, With Hints for Whom It May Concern,* by William Wright (New York, 1865); *Petroleum and Petroleum Wells: With a Complete Guide Book and Description of the Oil Regions of Pennsylvania, West Virginia and Ohio,* by J. H. A. Bone (Philadelphia, 1865); *Petrolia: A Brief History of the Pennsylvania Petroleum Region, Its Development, Growth, Resources, etc., from 1859 to 1869,* by Andrew Cone and Walter R. Johns (New York, 1870); *Petroleum: A History of the Oil Region of Venango County, Pennsylvania, Its Resources, Mode of Development, and Value; Embracing a Discussion of Ancient Oil Operations, etc.,* by S. J. M. Eaton (Philadelphia, 1866); *Derrick and Drill, or An Insight into the Discovery, Development, and Present Condition and Future Prospects of Petroleum in*

New York, Pennsylvania, Ohio, and West Virginia, by Edmund Morris (New York, 1865) ; *Old Times in Oildom,* by George W. Brown (Oil City, 1911) ; and *A Few Scraps, Oily and Otherwise,* by A. W. Smiley (Oil City, 1907) .

Of particular value on special phases of the development are the following: *The History of Pithole,* by "Crocus" (Charles C. Leonard) (Pithole City, 1867) ; *Coal Oil Johnny,* by John Washington Steele (Franklin, Pa., 1902) ; *The Life and Letters of James Abram Garfield,* by Theodore Clarke Smith (2 vols., New Haven, 1925) ; *Jay Cooke,* by Henrietta M. Larson (Cambridge, 1936) ; *Jay Cooke: Financier of the Civil War,* by Ellis Paxson Oberholtzer (Philadelphia, 1907) ; *Autobiography of Andrew Carnegie,* edited by John C. Van Dyke (New York, 1924) ; *A Practical Treatise on Coal, Petroleum, and Other Distilled Oils,* by Abraham Gesner (New York, 1861) ; *A Practical Treatise on Petroleum,* by Benjamin J. Crew (Philadelphia, 1887) ; *In French Creek Valley,* by John Earle Reynolds (Meadville, 1938) ; and *Old Time Tales of Warren County,* by Arch Bristow (Meadville, 1932) .

E. Dictionaries and Encyclopaedias

The Dictionary of American Biography and *Appleton's Cyclopaedia of American Biography* provide biographical data on a few of the early oilmen. Many of the prominent early oilmen, however, are not listed in either work.

F. Pamphlets

Especially scarce and valuable as the first account to be written and published after the completion of the Drake well is *The Wonder of the Nineteenth Century: Rock Oil in Pennsylvania and Elsewhere,* by Thomas Gale (Erie, 1860) . The depression of 1866 and 1867 in the oil field, the monopolies depressing the price of oil, and the situation of the producer are ably discussed in *A Crisis in the Oil Region: A Few Words in Behalf of the Producer,* by John Ponton (Titusville, 1867) . *New Haven and the First Oil Well,* by H. H. Townshend (New Haven, 1934) , is especially important in that it forcibly calls attention to the role played by the New Haven capitalists in drilling Drake's well at Titusville.

G. *Government Documents*

A comprehensive report on the history and development of the petroleum industry, the chemistry of petroleum, the production, transportation, and storage of petroleum, the technology of petroleum, and the uses of petroleum may be found in S. F. Peckham's "Production, Technology, and Uses of Petroleum and Its Products," *House Miscellaneous Document* No. 42, 47 Cong., 2 sess., part 10, 1–319. An extensive bibliography is included. An excellent but short account of the early beginnings of the petroleum industry may be found in Joseph C. G. Kennedy's "Preliminary Report on the Eighth Census, 1860," *House Executive Document* No. 116, 37 Cong., 2 sess., 71–75. The organization of the early companies, the spread of operations down Oil Creek, the speculation in oil, prices, refining, the export trade, and the effects of taxing petroleum are well reviewed in S. S. Hayes' "Report of the United States Revenue Commission on Petroleum as a Source of National Revenue," *House Executive Document* No. 51, 39 Cong., 1 sess., 1–39. For information concerning the export trade in petroleum, the reports of the Secretary of State on the commercial relations of the United States and of the Secretary of the Treasury on the commerce and navigation of the United States are indispensable.

H. *Bibliographies*

Petroleum and Natural Gas Bibliography, by Robert E. Hardwicke (Austin, Texas, 1937), has a section relating to the history of the petroleum industry, which should not be overlooked. The works by Peckham, and Arnold and Kemnitzer, cited above, also contain valuable bibliographical material.

INDEX

Abbott, W. H., amount of business transacted in 1869, 177; builds first refinery in Titusville, 91; buys interest in Barnsdall well, 68; buys interest in pipe line, 147

Accommodation pipe line at Pithole, 145

Actors and actresses in Titusville, 180

Albertite, 19, 22

Alden, John, xv, xvi

Alden, Timothy, xvi

Allegheny College, establishment of, xvi

Allegheny River, 10, 23, 74; effect on oil ring, 186–87; pipe lines to, 142, 144; transportation on, 15, 93, 132

Allegheny Transportation Co., first large pipe-line company, 147

Allegheny Valley R.R. Co., in oil ring, 186

Allen. See Hinkley & Allen, refiners

American Hotel (Titusville), Drake and his family live at, 53; headquarters of oilmen, 182; meeting of oilmen at, 190

"American Medicinal Oil, Burkesville, Ky.," 6

American Oil, 11, 24

Ames, F. W., hardware merchant, 177

Analysis of oil, 40, 96

Angier, J. D., abandons collection of oil, 46; leases land to collect oil, 30; ships oil to New York, 33

Ann Arbor, Michigan, Drake goes to, 48

Antwerp, Belgium, 96, 99

Appleton & Co., 33

Appleton's Collegiate Institute, Titusville, 179

Archbold, John D., leader in Culver Literary Association, 179; son-in-law of Major S. M. Mills, 182

Argand Burner, 40

Argyle, Pa., 192

Armstrong County, Pa., spread of oil operations to, 192

Atkinson farm, 164

Atlantic & Great Western R.R., 92, 96, 101, 111–13, 148–51, 154, 171

Atlantic & Pacific Telegraph Co., 182

Atwood, Luther, chemist, 20–22, 36, 92

Austro-Prussian War, effect on foreign market, 156

Babbitt of Babbitt's Schnapps, 58

Babylon, Pa., 157

Backus City, Pa., 165

Baku, oil in, 40

Baltimore Petroleum Co., establishes Oleopolis, 144

Bank failures, 154–55

Baptists in Titusville, 178

Barges, for transporting oil, 110

Barnsdall, William, builds first refinery in Titusville, 91; drills well, 68

Barnsdall Well, 68

Barnum, P. T., lectures in Titusville, 179

Barrows & Co., use first pipe line, 142

Bates, James A., buys Rooker farm, 131

Bean Well, 167

Beatty, Martin, drills salt well, 6

Belgium, marketing of oil in, 96

Bell, Edwin C., 197, 198

Bellomont, Earl of, early reference to oil, 1

Bennehoff, John, 161–63

Bennehoff, Joseph, 161, 162

Bennehoff City, 160

Bennehoff Petroleum Co., 147

Bennehoff robbery, 162

Bennehoff Run, 146, 158, 159, 161

Bibliography of period, 197–203

Billings, Josh, lectures in Titusville, 179

Binney, E. W., 19

INDEX

Petroleum Reporter, purchased by Bloss brothers, 173
Petrolia, Pa., 192; Oil Exchange, 191
Phil Sheridan Well, 159
Philadelphia, Pa., 23, 25; in oil ring, 184; J. W. Steele in, 116–17
Philadelphia & Erie R.R., 92, 101, 111–12, 142, 148–51, 149
Philadelphia Ledger, on demand for oil stocks, 122
Philadelphia Well, 159
Phillips, William, 80
Phillips No. 1 Well, 80
Phillips No. 2 Well, 80, 87
Philpot, Brian, 133, 146
Photographs, in Drake Museum, 198
Pickett, Henry, 133, 146
Pierpont, Asahel, 42–43, 44, 46
Pine Creek, 6, 30
Pioneer City, Pa., 161
Pioneer Run, 160
Pioneer Wells, 62–75
Pipe lines, 141–49; Abbott & Harley, 147; accommodation pipe line, 145; Allegheny Transportation Co., 147; capacity of, 144, 146; construction of, 143–44; cost of, 146; difficulties in construction, 142; Henry Harley, 146–47; opposition to, 141–42, 143; Pennsylvania Tubing & Transportation Co., 144; Titusville Pipe Co., 145–46, 149; to Henry's Bend, 144; Van Syckel, 143–44, 147
Pithole City, 127–40; buildings, 136; buildings moved to Shamburg, 165; churches, 138; civic improvements, 139; decline, 140; growth in population, 139; hotels, 139; laying out town, 135; map, 129; pipe lines from, 143, 144; production of oil, 135; railroads, 149–50; social conditions, 136–39; theater, 138; topography, 135
Pithole Creek, 127, 144; production of oil, 135, 140
"Pithole's Forty Thieves," 138–39
Pits, early source of oil, 6–9
Pittsburgh, Pa., 23, 24, 26; in oil ring, 184; Oil Exchange, 191; oil transported to, 109; petroleum introduced in, 15; refineries, 25, 93
Pittsburgh & Cherry Run Petroleum Co., 164
Pleasantville, Pa., 54, 164; discovery of oil, 165–66; social conditions, 167
Plumb, Capt. Ralph, 124

Plumer, Pa., 92, 129, 142, 149
Pond freshets, 103–9
Ponton, John, xxxiv
Pool Well, 134
Porter farm, 165
Portland Kerosene Co., 27
Pouchot, mentions oil, 1
Pownall, Thomas, reference to *Map of the British Colonies in North America* in 1776, 13
Prather, George C., buys Holmden farm, 128; refuses to sell Holmden farm, 134
Pratt, F. M., 132
Presbyterians in Titusville, 178
Prices. *See* Oil, price of; Farms, price of; Transportation charges
Price-fixing, 84
Production. *See* Oil Wells, production of
Production of oil:
  in 1859, 75
  in 1860, 75
  in 1861, 83
  in 1862, 85
  in 1863, 87
  in 1864, 121
  in 1865, 131, 135, 140
  in 1866, 154, 157
  in 1867, 163
  in 1868, 165, 167
  in 1869, 192
Profits. *See* Dividends; Oil Wells, income from; Oil, price of
Promotion in oil industry, 27, 37, 42, 46, 88–100

"R. C. T.," motto of Swordsman's Club, 138
Railroads, 149–51; Atlantic & Great Western, 92, 96, 101, 111–13, 148–51, 154, 171; Farmers', 49; Oil Creek, 111, 117, 143, 150, 160, 172, 177; Oil Creek & Allegheny River, 150–51; Oleopolis & Pithole, 149; Pennsylvania, 148; Philadelphia & Erie, 92, 101, 111–12, 142, 148–51; Reno & Pithole, 150, 154; Warren & Franklin, 149–51. *See also* Transportation, by railroad
Red Fox, steamer on Allegheny River, 109–16
Reed, James, 141
Reed, William, 119
Reed Well, 119
Refineries, 25, 91–95, 193; Downer, 92–93; Humboldt, 127, 142

Use and Abuse

of

America's Natural Resources

An Arno Press Collection

Ayres, Quincy Claude. **Soil Erosion and Its Control.** 1936

Barger, Harold and Sam H. Schurr. **The Mining Industries, 1899–1939.** 1944

Carman, Harry J., editor. **Jesse Buel:** Agricultural Reformer. 1947

Circular from the General Land Office Showing the Manner of Proceeding to Obtain Title to Public Lands. 1899

Fernow, Bernhard E. **Economics of Forestry.** 1902

Gannett, Henry, editor. **Report of the National Conservation Commission, February 1909.** Three volumes. 1909

Giddens, Paul H. **The Birth of the Oil Industry.** 1938

Greeley, William B. **Forests and Men.** 1951

Hornaday, William T. **Wild Life Conservation in Theory and Practice.** 1914

Ise, John. **The United States Forest Policy.** 1920

Ise, John. **The United States Oil Policy.** 1928

James, Harlean. **Romance of the National Parks.** 1939

Kemper, J. P. **Rebellious River.** 1949

Kinney, J. P **The Development of Forest Law in America.** *Including,* Forest Legislation in America Prior to March 4, 1789. 1917

Larson, Agnes M. **History of the White Pine Industry in Minnesota.** 1949

Liebig, Justus, von. **The Natural Lawss of Husbandry.** 1863

Lindley, Curtis H. **A Treatise on the American Law Relating to Mines and Mineral Lands.** Two volumes. 2nd edition. 1903

Lokken, Roscoe L. **Iowa**—Public Land Disposal. 1942

McGee, W. J., editor. **Proceedings of a Conference of Governors in the White House, May 13–15, 1908.** 1909

Mead, Elwood. **Irrigation Institutions.** 1903

Moreell, Ben. **Our Nation's Water Resources**—Policies and Politics. 1956

Murphy, Blakely M., editor. **Conservation of Oil & Gas: A Legal History, 1948.** 1949

Newell, Frederick Haynes. **Water Resources:** Present and Future Uses. 1920.

Nimmo, Joseph, Jr. **Report in Regard to the Range and Ranch Cattle Business of the United States.** 1885

Nixon, Edgar B., editor. **Franklin D. Roosevelt & Conservation, 1911–1945.** Two volumes. 1957

Peffer, E. Louise. **The Closing of the Public Domain.** 1951

Preliminary Report of the Inland Waterways Commission. 60th Congress, 1st Session, Senate Document No. 325. 1908

Puter, S. A. D. & Horace Stevens. **Looters of the Public Domain.** 1908

Record, Samuel J. & Robert W. Hess. **Timbers of the New World.** 1943

Report of the Public Lands Commission, with Appendix. 58th Congress, 3d Session, Senate Document No. 189. 1905

Report of the Public Lands Commission, Created by the Act of March 3, 1879. 46th Congress, 2d Session, House of Representatives Ex. Doc. No. 46. 1880

Resources for Freedom: A Report to the President by The President's Materials Policy Commission, Volumes I and IV. 1952. Two volumes in one.

Schoolcraft, Henry R. **A View of the Lead Mines of Missouri.** 1819

Supplementary Report of the Land Planning Committee to the National Resources Board, 1935–1942

Thompson, John Giffin. **The Rise and Decline of the Wheat Growing Industry in Wisconsin** (Reprinted from *Bulletin of the University of Wisconsin,* No. 292). 1909

Timmons, John F. & William G. Murray, editors. **Land Problems and Policies.** 1950

U.S. Department of Agriculture—Forest Service. **Timber Resources for America's Future:** Forest Resource Report No. 14. 1958

U.S. Department of Agriculture—Soil Conservation Service and Forest Service. **Headwaters Control and Use.** 1937

U.S. Department of Commerce and Labor—Bureau of Corporations. **The Lumber Industry,** Parts I, II, & III. 1913/1914

U.S. Department of the Interior. **Hearings before the Secretary of the Interior on Leasing of Oil Lands.** 1906

Whitaker, J. Russell & Edward A. Ackerman. **American Resources:** Their Management and Conservation. 1951